$17.25

Estimating Plumbing Costs

By Howard C. Massey

Craftsman Book Company
6058 Corte del Cedro, P.O. Box 6500
Carlsbad, CA 92008

To my wife, Hilda, for her untiring assistance in editing and typing this manuscript.

Copyright 1982 Craftsman Book Company

Library of Congress Cataloging in Publication Data

Massey, Howard C.
    Estimating plumbing costs.

    Includes index.
    1. Plumbing--Estimates. I. Title.
TH6235.M33   696'.1   82-5010
ISBN 0-910460-82-5   AACR2

Edited by Sam Adrezin

Second Printing 1984

Contents

Introduction..............................5

Part 1 Interpreting Plans & Drawings..........6

1 Beginning the Estimate....................7
    Plot Plans...............................8
    What a Plot Plan Shows..................8

2 Floor Plans and Layouts..................14
    Floor Plans for the Handicapped........14
    Minimum Plumbing Fixture Clearances....15
    Sample Architect Floor Plans...........17
    Commercial Plot Plan...................30
    Commercial Isometric Drawing...........32

3 Isometric Drawings......................35
    How to Make Isometric Drawings.........35
    Fittings Within an Isometric Drawing...38
    Classes of Pipes and Fittings..........41

Part 2 Special Plumbing Systems.............42

4 Interceptors and Special Waste Piping.....43
    Grease Interceptors....................43
    Gasoline, Oil and Sand Interceptors....44
    Slaughter House Interceptors...........45
    Laundry Interceptors...................45
    Other Trap Requirements................46
    Dilution or Neutralizing Tanks.........47
    Special Waste Piping...................48
    Storm Drainage Systems.................49
    Storm Rainfall Planning Requirements...51

5 Trailer Park Systems....................55
    Trailer Park Sanitary Facilities.......55
    Sizing Park Drainage Systems...........56
    Estimating Trailer Park Systems........57
    Water Distribution Systems.............58

6 Fire Protection Systems.................59
    Standpipe Systems......................59
    Estimating Material and Installation...59

Typical Connection Diagram.............60
Fire Pump Detail.......................61
Thrust Blocks..........................62
Standpipe Layout and Connections.......63
Water Sources and Pumps................65
Fire Wells and Connections.............66
Yard and Street Hydrants...............68

7 Gas Systems.............................69
    Kinds of Gas...........................69
    Sizing Gas Systems.....................70
    Commercial Installation................71
    Residential Installation...............72
    Estimating Gas Systems.................73

8 Sanitary Sitework and Water Distribution Systems..........75
    Sanitary Sitework Systems..............75
    Private and Public Sewer Connections...75
    Sewage Lift Stations, Ejectors and Sump Pumps...76
    Standard Sanitary Sewer Detail.........78
    Sewage Lift Stations...................88
    Emergency Connection Riser Detail......90
    Force Main Connection..................91
    Transformer Vault Drainage.............93
    Trash Chute Piping.....................94
    Transformer Holding Tank...............95
    Trap Reseal Detail.....................96
    Settling Tank and Drainage Well Detail...97
    Accessory Building Hook-up.............99
    Wet Vent System.......................101
    Water Distribution System.............105

Part 3 The Estimating Process.............110

9 The Contractor and His Estimator........111
    Importance of Accurate Bidding........111
    Bidding Limitations...................111
    Good Job Management...................111
    Know Your General Contractor..........112

10 The Plumbing Estimator.................113
Common Mistakes.....................113
Components of a Good Estimate........113
What Causes Business Failure..........114
The Detailed Estimate...............114
Office Overhead Expense...............114
"Cost Plus" Jobs.....................115
Job Overhead Expense................115
Special Expenses......................115

11 The Subcontract Agreement............117
Advance Considerations...............117
Need for Clarity and Precision........117
Standard Subcontract Agreement........118
Elements of a Contract...............119
Compensations........................119
Penalties and Indemnity Clauses........119
"Cost Plus" Contracts................119
Contract Cancellations...............119
Completion Problems..................120
Breach of Contract....................120
Importance of Written Specifications....120
Read the Fine Print..................121
Faulty Work and the Courts...........121
Extra Charges........................121
Unexpected Conditions................121
Definition of Substantial Performance...121
General and Subcontractor Relationship..122
Lien Laws............................122

12 The Material Take-off..................123
Specifications and Blueprints..........123
General Estimating Outline............124
Using Man-hour Tables................124

13 Estimating Forms.....................126
Quantity Cost Sheets..................126
Standard Analysis Sheet...............127
Quantity Cost Sheet Form..............128
Labor Cost Summary Sheet Form.......130
Project Summary Sheet Form..........131

14 Taking-off & Pricing the Estimate........133
Procedure for Figuring Labor Cost......133
Drainage, Waste, and Vent System......133
Piping Quantity Cost Sheet............135
Fitting Quantity Cost Sheet...........136
Other Drainage Systems...............137
Sample Estimates.....................137
Consideration of Material Discounts.....140
Quantity Cost Sheets.................141
Labor Cost Summary Sheets..........146
General Summary Check..............149

15 Completing the Estimate................155
Forwarding Costs to Project
 Summary Sheet.....................155
Adding Supplementary Costs..........155
Project Summary Sheet, Sample........156
Contract Based on Estimate............157

16 Man-hour Tables for Plumbing..........159
An Average Plumber's Hour...........159
Labor Cost...........................159
How to Use the Tables................159
Building Piping Systems..............159
Man-hour Tables.....................160
Converting Minutes to Decimal
 Hours, Table........................169
Horizontal & Vertical Piping Supports...170

Part 4 Appendix........................172

A Job Specifications......................173
Typical Specifications for Simple Job....173
Typical Specifications for Large
 Commercial Building.................174

B Calculating Water Supply &
Fixture Requirements 183
Calculating Water Supply Systems......183
Method One.........................185
Method Two.........................186
Probable Demand.....................187
Maximum Fixture Demand............188
Calculating Fixture Requirements and
 Sewage Flow.......................190
Types of Establishments..............191

C Making Plumbing Calculations..........197
Circle Measurements.................197
Piping Offsets & Pipe Equivalents.......198
Water Pressure and Flow Rate.........200
Fixture Pipe Sizes....................201

D Definitions 203

E Standard Abbreviations..................211

F Common Fixture, Plumbing,
& Fitting Symbols....................213
Fixture Symbols.....................213
Plumbing Symbol Legend.............216
Fitting Symbols.....................217

Index..................................220

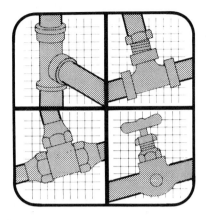

Introduction

Part 1
Interpreting Plans and Drawings

Part 1 of this book is essential to every estimator who expects to compile accurate plumbing costs.

Chapter One discusses why estimators need to be familiar with plot plans and how to interpret them. It explains why the estimator should visit the job site and what to look for when there.

Chapter Two discusses floor plans as an essential part of all construction blueprints. It will refresh your memory on plumbing for the handicapped and help you design drainage, waste and vent systems as depicted on architectural prints.

Chapter Three focuses on isometric drawings as an essential part of the plumbing estimator's work. They are used by plumbing contractors to design piping layouts, indicate the number and type of fittings, and calculate labor for installation.

Part 2
Special Plumbing Systems

Part 2, Chapters Four through Eight, discusses special piping systems and their components. It contains illustrations of interceptors and special waste piping, trailer park, fire protection and gas systems.

Part 2 has been arranged to bring together design objectives and code requirements for the estimator to use as a ready reference on jobs he is bidding.

Part 3
The Estimating Process

Part 3, Chapters Nine through Sixteen, deals exclusively with the process of estimating. The responsibilities and qualifications of the plumbing estimator are covered, as are subcontract agreements, estimating forms and comprehensive man-hour production tables. Step-by-step instructions show how to take off materials and labor, and how to price a plumbing job.

This section presents sound estimating procedures that will work well on any plumbing job.

Part 4
Appendix

Appendix A consists of sample job specifications for residential and commercial jobs. It shows how the architect and engineer spell out very detailed requirements for roughing-in materials, fixtures and even installation methods.

Appendix B provides two methods an estimator may use to calculate water supply systems, and shows why he needs to know how to make these calculations when working with architects in designing drainage systems.

Appendix C supplies equations and tables to help the estimator make the necessary calculations, and shows, with examples, how to use them.

Appendix D provides the definitions found in most plumbing codes.

Appendix E gives abbreviations used on blueprints and plumbing reference books (including most codes) to identify plumbing fixtures, pipes, valves and nationally-recognized associations.

Appendix F gives common fixture symbols, plumbing symbol legends and fitting symbols.

Part 1

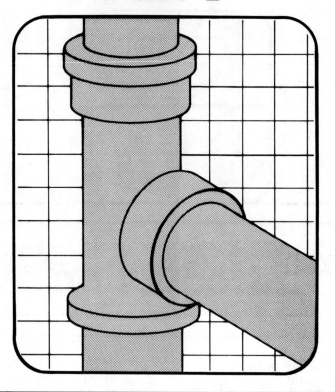

Interpreting Plans & Drawings

- **Beginning The Estimate**
- **Floor Plans & Layouts**
- **Isometric Drawings**

1

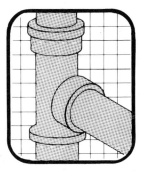

Beginning The Estimate

Most plumbers who go into contracting today are exceptional mechanics, foremen or superintendents. They have a good understanding of the plumbing code and know how the work should be done. But most lack experience in business management and have very little, if any, training in estimating job costs.

Unfortunately, few colleges, technical schools or vocational schools teach plumbing estimating. On-the-job experience with a successful contractor is usually the only training available for anyone who wants to master the fundamentals of estimating plumbing work. Unfortunately, this can be a slow and expensive way to learn any skill. Some inexperienced estimators stay in business making a reasonable profit. Others fail and eventually go back to working as tradesmen.

But it is true that good estimating is more important today than ever. "Normal times" for the construction industry no longer exist. Inflation, shortages, and government policy intrude on every business, large or small. Labor requirements have become more difficult to estimate. Add to these problems the daily aggravations of personnel and financial burdens and you can see why every plumbing estimator needs the best possible training for the work he must handle.

The main focus of this book is to set up a basic procedure that will work for every plumbing estimator. This book won't automatically relieve you of responsibility for determining material and labor costs. But it will be an excellent guide when estimating the materials, labor, direct and indirect costs in every plumbing job.

Materials and installation methods are *not* included in this book. *Plumber's Handbook* by this author provides in-depth information on sizing plumbing systems, plumbing materials and installation methods and will be a valuable aid to every plumbing estimator. But this book does include extensive man-hour tables compiled by the author and other reference materials that will be essential when compiling any estimate.

Every estimate begins with knowledge of what the job requires. That means you must understand what the plumbing code mandates for the job you are bidding. If your plumbing job won't pass inspection, you aren't going to get paid. And most owners will be very reluctant to pay for anything required by the plumbing code but not included in your estimate. So as an estimator you are charged with estimating a plumbing system that *will* pass inspection, even if the plans and specifications you bid from don't show what the plumbing code requires.

Because estimators have to base their estimates on code requirements, the first eight chapters emphasize points you may overlook when developing your list of materials.

The next two chapters examine the contractor's and estimator's role in the construction process and detail key points that make a contractor or estimator more successful than others in his profession.

The last four chapters deal with the nuts and bolts of the estimating procedure: the material take-off, pricing, estimating checklists and sample estimates. Man-hour tables are included in this section. An appendix at the back of this manual includes reference data for your use when compiling estimates.

Every estimator should learn to follow established routines, work systematically and follow a logical sequence to avoid errors. In the logical sequence of estimating, the plot plan comes first. So that is where we will begin.

Plot Plans
The drawing of a plot of ground on which a building will be constructed is known as a *plot plan*. Since most sets of plans include a plot plan, you should know how to interpret them. A plot plan shows:
- The shape and size of the lot.
- The setbacks from the property lines.
- All permanent outside construction above ground, including the driveways, easements (if any), streets and the electric service, as well as other valuable information.
- All below-ground information including the source, location and size of the water facility; the location and size of the sewage facilities; the location and size of soakage pits and catch basins; the type, location and size of the storm water drainage piping, and the location and size of fire protection systems.

Plot plans are required by code and are an essential part of all construction plans. *Residential* plot plans are usually drawn to scale so that 1'' equals 20'0''. *Commercial* plot plans are usually drawn to a scale of 1'' to 40'0''. Plot plans are first approved by a plans examiner. They are then called "as-built" plans.

What is shown on the plans and approved by the plans examiner must be strictly adhered to. Plumbing estimators should check the approved job set of plans at the general contractor's office to see if there are any special requirements. Making changes to the plans after approval requires submitting a revised plot plan for reapproval. This is an extra expense for the plumbing contractor.

Plan to visit the job site for the building you are estimating, if possible. The site may be low, steep or rocky, requiring extra equipment or labor to install the plumbing site-work. Check with the utility company for sewer lateral location and depth as well as the water meter location and frost line depth.

If public sewers are not available and a septic tank is required, and if your company intends to install the septic tank, check with the local Health Department or the Department of Environmental Resource Management (DERM) for any special requirements. Check with these agencies if a domestic well is to be used where public water is not available.

Most plumbing contractors do not install septic tanks or wells. Job specifications sometimes require the plumbing contractor to assume this responsibility and include these prices in his contract. When this is true, contact several reliable specialty contractors and have each one submit bids for this portion of the contract. Select one, include the sub-contractor's bid, plus your own profit on the sub's job cost, and add it to the contract. The plumbing contractor is entitled to and should make a profit on all work done by subcontractors.

The plot plans presented in Figures 1-1 through 1-4 show the four basic types of outside plumbing facilities: (1) a residential building having public water and sewer, (2) a residential building having public water and septic tank, (3) a residential building having a well and a septic tank, and (4) a commercial building having public water, sewer and a soakage pit.

The house in Figure 1-1 is a typical single-family dwelling. It is connected to a public sewer located at approximately the center of the lot. The water service is connected to a water meter located on the right side of the lot. In this plan the utilities servicing the dwelling are in the street and not in the utility easement at the rear of the property.

The plot plan in Figure 1-2 shows a house in a suburban area. It has public water, but depends on a septic tank for sewage disposal. The rapid growth of urban communities around large metropolitan areas has often exceeded the capacity of sanitary sewage collection systems. Check with the local utility company to determine if sewer hook-ups are available. Don't assume that the architect's plans are right unless they have been approved by the local building and zoning department. On site sewage disposal involves extra costs that could reduce or eliminate the plumbing contractor's profit on a small job.

The house in Figure 1-3 is in a rural area. It depends on a private well for water and on a septic tank for sewage disposal. On small jobs such as this, the general contractor will usually sub-let the well and septic tank to specialty contractors. If so, the estimator will include in his bid only the cost of the house water connection to the hydropneumatic tank and the house building drain to the septic tank.

Beginning The Estimate

The commercial building in Figure 1-4 has public water and sewer connections. The estimator will estimate the cost of connecting the building to the public sewer located at the property line. Check with the utility company for exact location and depth of the sewer line. Also include the cost of connecting the building to the water meter located at the property line. Water meters are expensive.

Check job specifications or with the builder to determine whether the water meter is included in the plumbing work.

The soakage pit and catch basins are designed to dispose of surface water from a parking lot. They are usually not considered to be part of plumbing work.

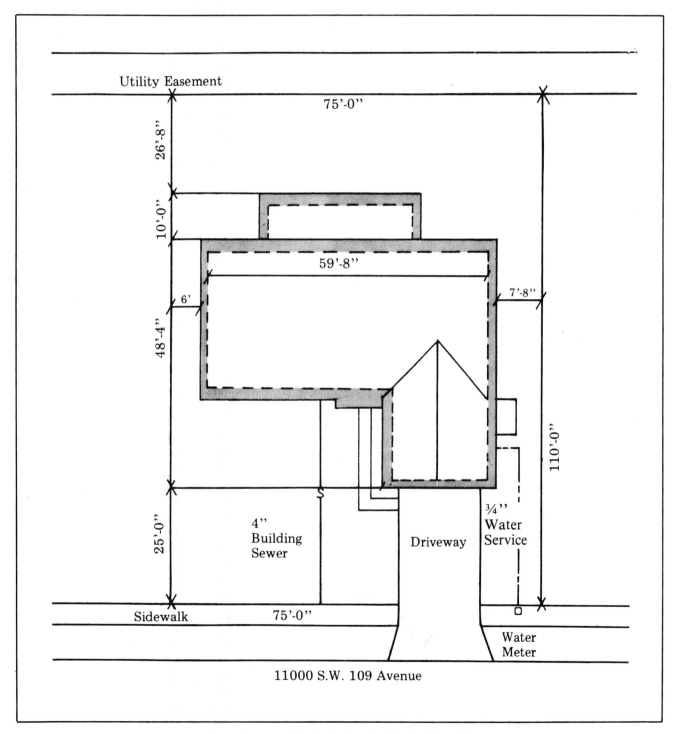

Plot Plan — Typical Single Family Dwelling
Figure 1-1

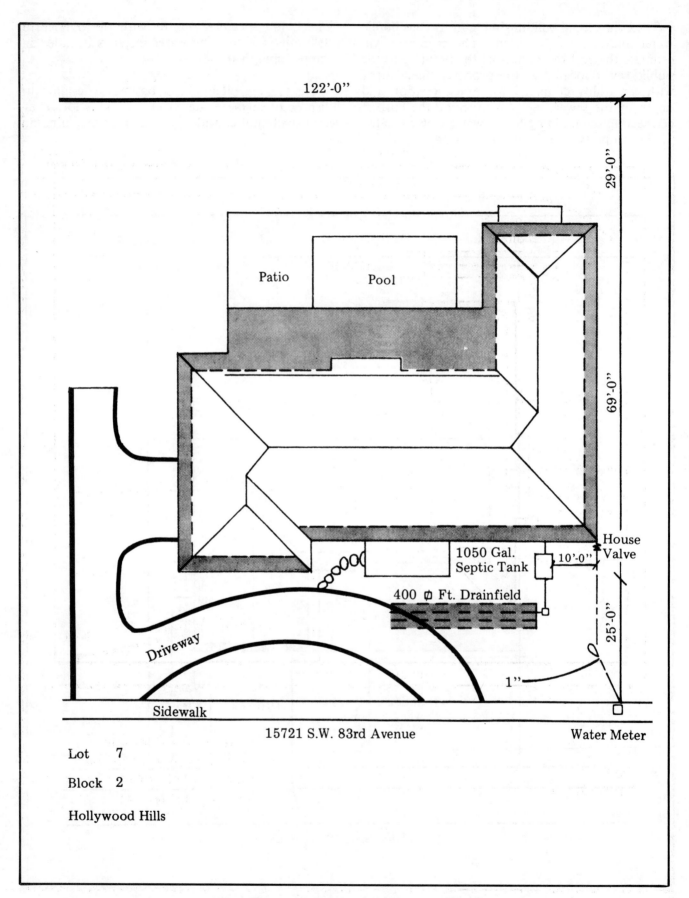

Plot Plan — Typical Suburban Dwelling
Figure 1-2

Beginning The Estimate

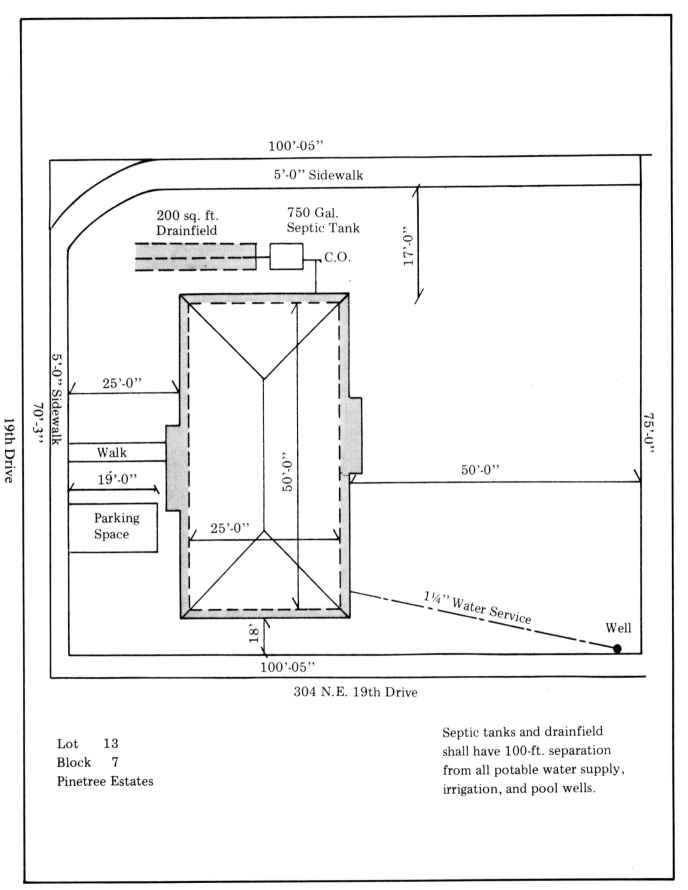

Plot Plan — Typical Rural Dwelling
Figure 1-3

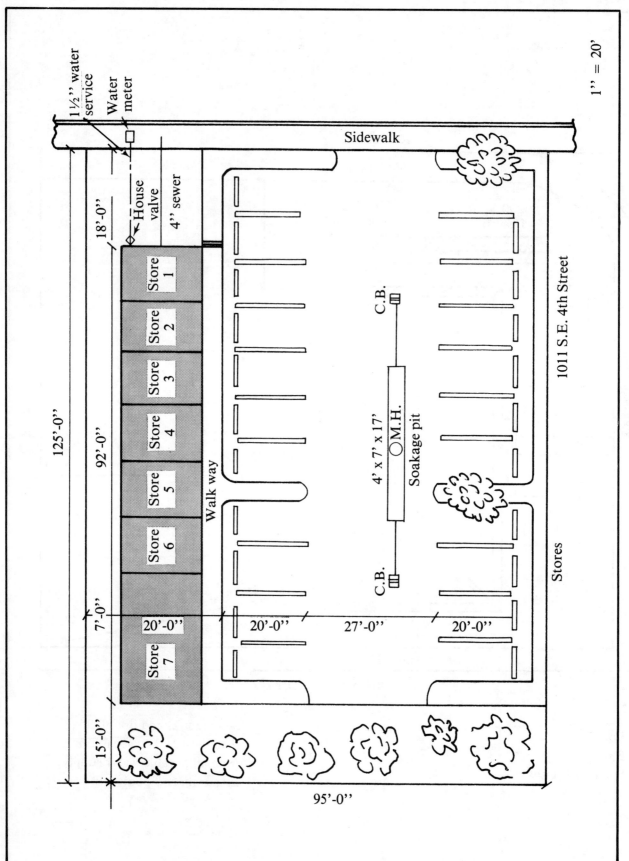

**Commercial Plot Plan
Figure 1-4**

2

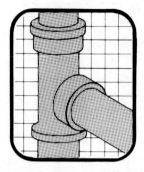

Floor Plans & Layouts

Layouts

This chapter shows examples of acceptable drainage, waste and vent systems as drawn from architectural plans. Plumbing details vary from code to code, but the basic principles of sanitation and safety are the same. What you find in this chapter should be applicable almost anywhere in the United States.

Floor Plans

Floor plans are required by code and are an essential part of all construction blueprints. The floor plan shows type and location of plumbing fixtures, floor dimensions, walls, doors, windows and other construction details.

The type and number of plumbing fixtures required by most model codes depends on the type of occupancy and the number of people using the facilities. Codes vary considerably so you will need to refer to the local code for exact requirements if you are bidding a job from unapproved plans.

Correct spacing of the plumbing fixtures is essential. Some bathrooms are designed too small to accommodate the specified fixtures. This should be brought to the attention of the architect. The plumbing contractor must space each fixture according to code requirements. Spacing must allow room for use, cleaning, and repair. Figure 2-1 shows minimum plumbing fixture clearances.

Table 2-2 gives the abbreviated symbols used to identify types of plumbing fixtures shown on floor plans and in isometric drawings. To designate a lavatory, architects may use either the letter "L," or the abbreviated word "Lav." These abbreviations will be used throughout this book. Completed abbreviations and fixture symbols are in Appendices E and F.

Floor Plans for the Handicapped

A plumbing estimator must know about fixtures required for the handicapped.

Federal and state laws now require that public and private buildings have toilet facilities available for the physically handicapped. Only single-family residences, apartment buildings and buildings considered hazardous are exempt from this requirement. The law states that all *required* restrooms in a building must be accessible by the handicapped. In large multi-story buildings, men's and women's facilities must be provided on each floor. In each *required* restroom, one toilet must comply with standards created by the President's Committee on Employment of the Physically Handicapped and by the American Standard Association. These standards, which are seldom included in construction blueprints, the plumbing code or in other reference books, include:

1. Toilet rooms must have at least one water closet that:

• has the seat 20 inches from the floor.

• has handrails, one on each side 33 inches high, and parallel to the floor. Handrails should allow a 1½ inch clearance between the rail and the wall.

Handrails must be securely fastened at each end and at the center.

Floor Plans & Layouts

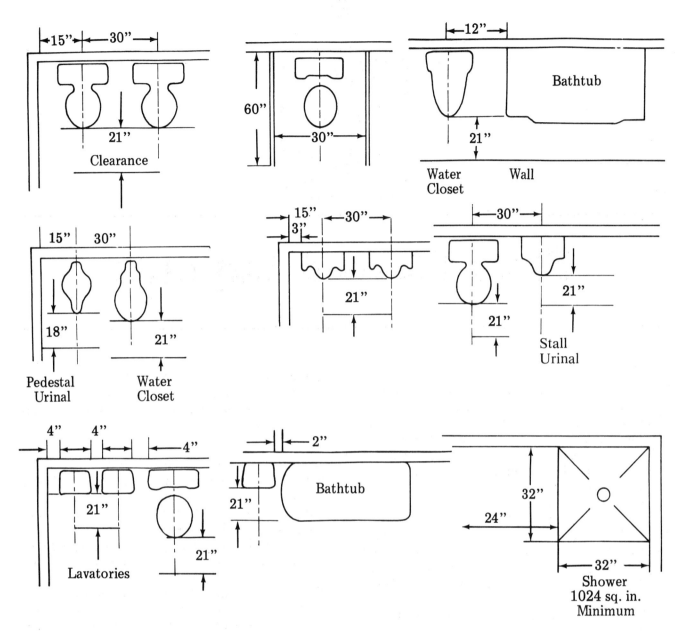

Minimum Plumbing Fixture Clearances
Figure 2-1

Plumbing Fixtures	Abbreviated Symbols
Water closet	W.C.
Bathtub	B.T.
Shower	Sh.
Lavatory	L
Kitchen Sink	K.S.
Clothes washing machine	C.W.M.
Water heater	W.H.

Plumbing Fixtures Abbreviated Symbols
Table 2-2

2. Toilet rooms must have at least one lavatory that:

• has a narrow apron which, when mounted at standard height, is usable by individuals in wheelchairs. It may be mounted higher, when the particular design demands, if it is usable by individuals in wheelchairs.

• has the drain pipes and hot water pipes insulated or located so that a handicapped individual without sensation will not injure himself.

3. Toilet rooms for men must have at least one urinal that:

• is floor mounted and level with the main floor

of the toilet room.

• is wall mounted with the opening of the basin no higher than 19 inches from the toilet room floor.

4. An appropriate number of water coolers or fountains must be accessible and usable by the physically handicapped. These are to be hand operated or hand and foot operated and are to have up-front spouts and controls.

Wall-mounted, hand operated coolers 36 inches from the floor can be used by both able bodied and physically handicapped individuals.

Water fountains that are fully recessed are not acceptable for use by the handicapped. However, they can be set in an alcove wide enough (minimum 32 inches) to accommodate a wheelchair.

Start your plumbing design by drawing an isometric layout of the drainage and vent system. (Chapter Three reviews the basics of isometric drawings.) Fixture location and type must comply with the architect's floor plan. This will also be true when preparing your material take-off. Plan the layout carefully so that waste, vent or water piping are not installed in door or window openings or through obstructed wall, floor or ceiling cavities.

Make sure your layout uses the type of installation best suited for the job. Your choice may include the stacked vent system, the stacked wet vent system, the flat wet vent system, the individual vent system, or a combination waste and vent system. Your isometric drawing should indicate the type and number of fittings required. These will be listed and priced on the *material quantity cost sheet* (see Chapter Thirteen) and ordered from your supplier.

There are several acceptable ways to lay out the drainage, waste and vent systems on any job. Very few estimators will lay out the same job in exactly the same way. But it is important to make your layout as simple as possible. This saves material and labor and keeps your company competitive.

At this point you should familiarize yourself with the sample drawings on the following pages. Figures 2-3 through 2-9 show differing layouts of bathroom floor plans. Each figure has two isometric drawings illustrating two ways the rough plumbing may be installed. Floor plans and isometric drawings for four single-family residences are shown in Figures 2-10 through 2-13.

For a more challenging layout, take a look at the eight-unit apartment building shown in Figure 2-14. This shows the plot plan, the water and sewage location, and identifies the type of units the building contains. Figure 2-15 shows the first story floor plan of Apartment "C." Figure 2-16 shows the second story floor plan of Apartment "C." This is all the information you need to begin your plumbing layout.

Although Apartment "C" has been selected to illustrate the plumbing layout, Apartments A, B, and D would follow the same procedure.

Begin with the building collection system. Figure 2-17 shows the partial underground sanitary collection system. This includes the base of stacks A, E and F that will serve Apartment "C." Figure 2-18 shows the completed stacks extended through the roof. In larger buildings like this the stacks are drawn separately from the collection system and are lettered to correspond with positions in the collection system.

The final illustrations show a two-story commercial building with a restaurant. Figure 2-19 is the plot plan showing water supply lines, sewer, grease interceptor, soakage pit and catch basin locations. Figure 2-20 shows the floor plan of the two stores and the restaurant, and indicates the location and type of plumbing fixtures and equipment used. Figure 2-21 is an isometric drawing of the complete drainage, waste and vent system. Note that the commercial food grinder connects to the sanitary portion of the drainage system. Most codes prohibit connection to the greasy waste system. The grease interceptor is not designed for large amounts of solid waste which would overload it.

This chapter includes drawings of only the drainage, waste and vent systems. But these involve the most misunderstood and misinterpreted sections of the code. Master these drawings so you know why each line is run as indicated and you will be able to design most simple plumbing systems.

Floor Plans & Layouts

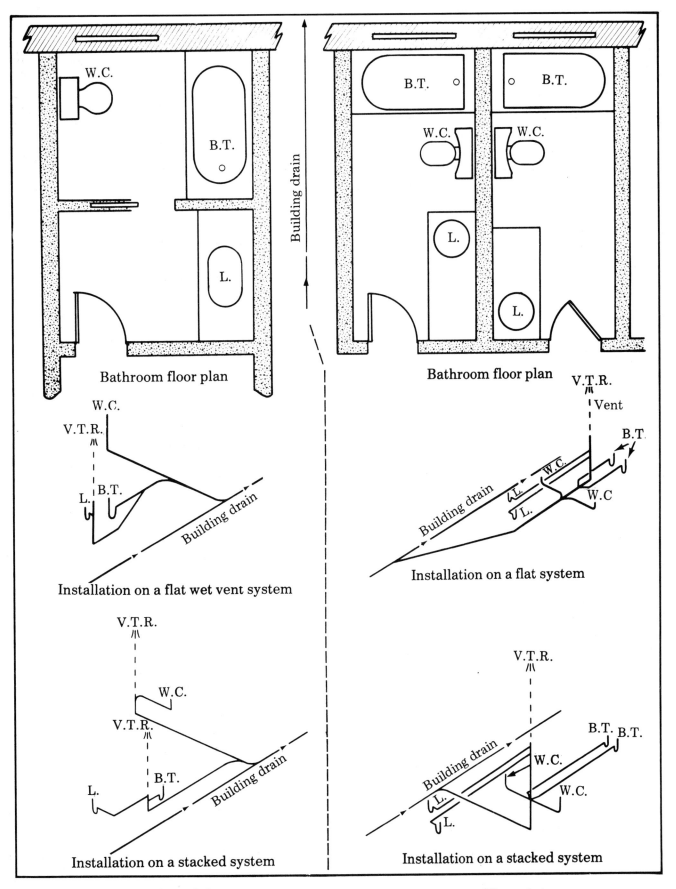

Figure 2-3   Figure 2-4

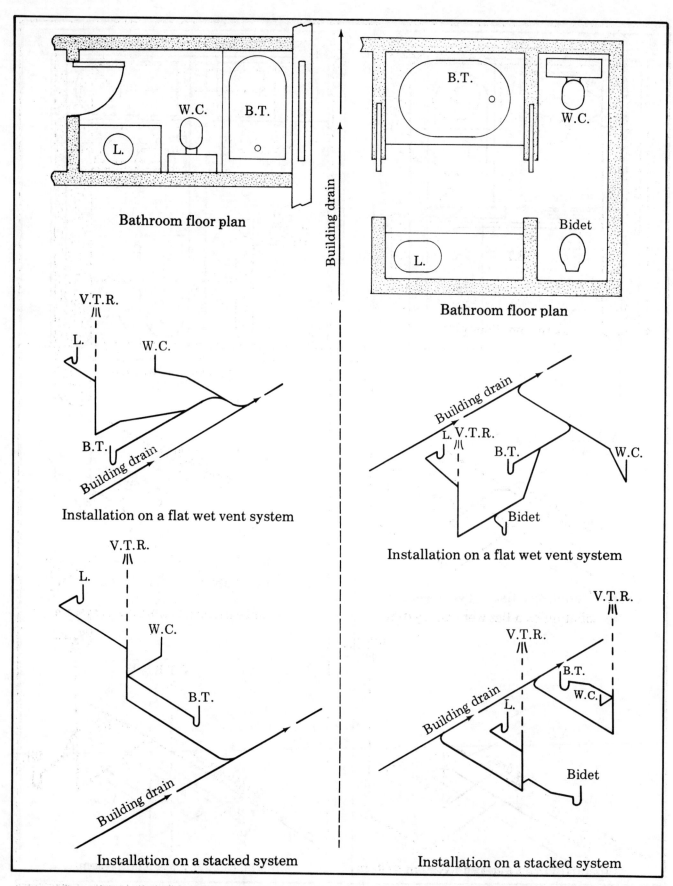

Figure 2-5

Figure 2-6

Floor Plans & Layouts

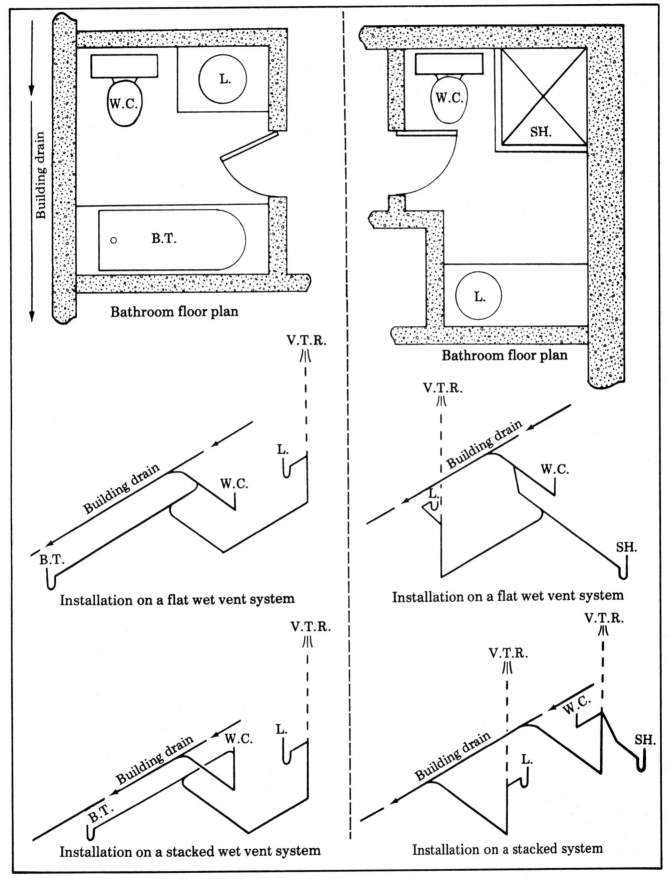

Figure 2-7

Figure 2-8

20 Estimating Plumbing Costs

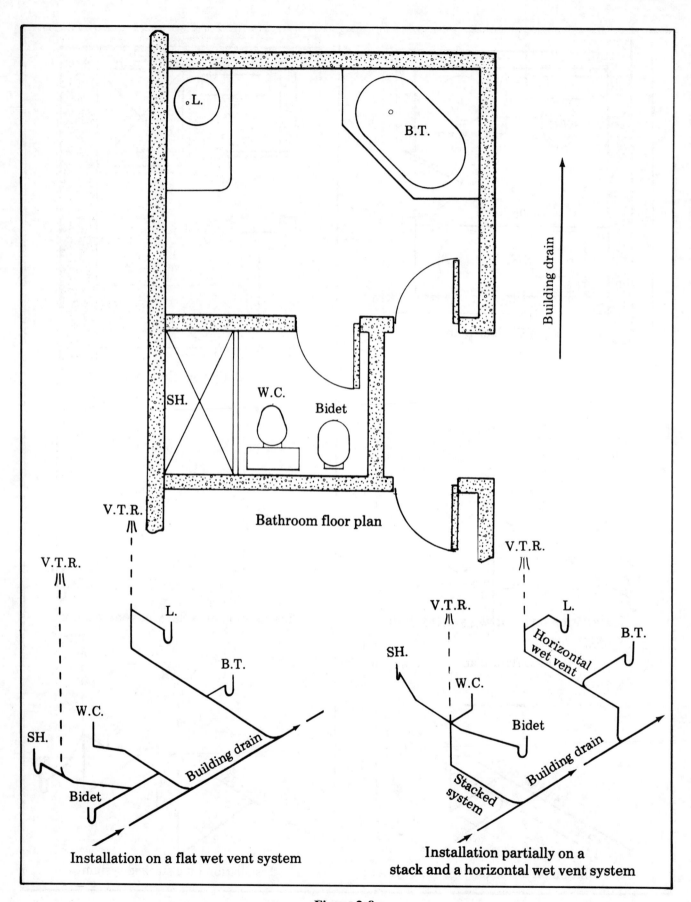

Figure 2-9

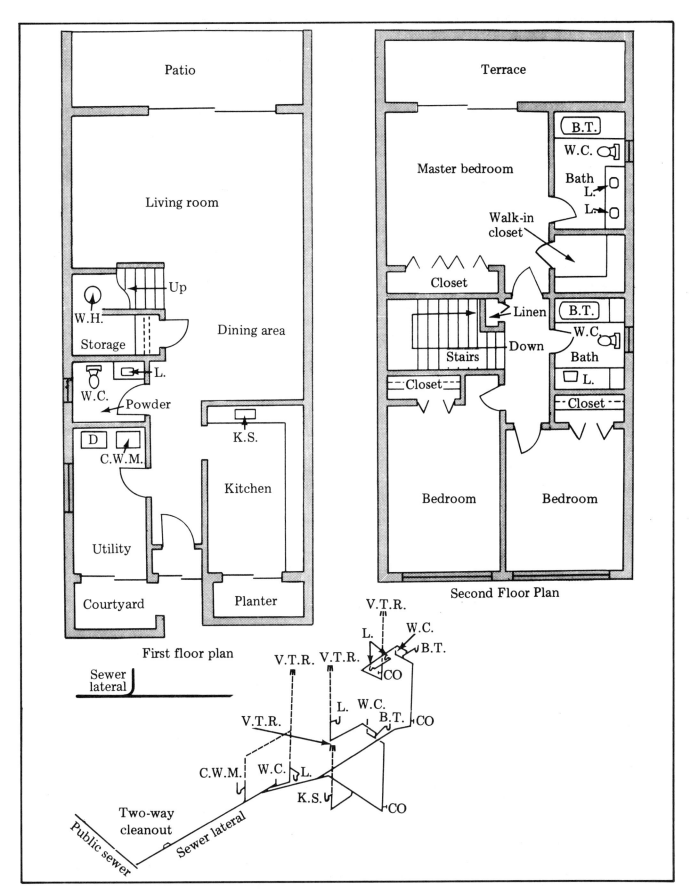

Single Family Residence Layout
Figure 2-10

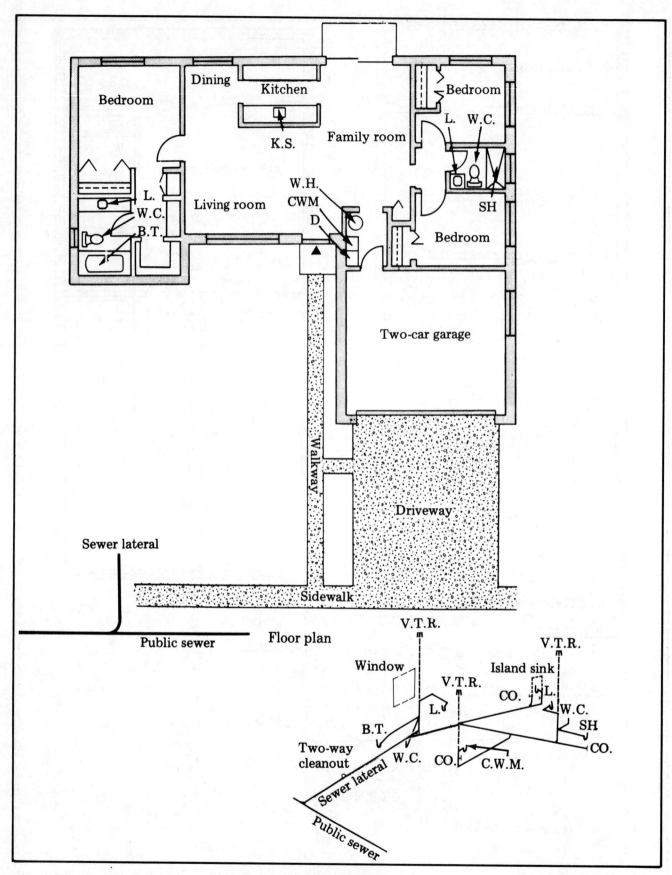

Single Family Residence Layout
Figure 2-11

Floor Plans & Layouts

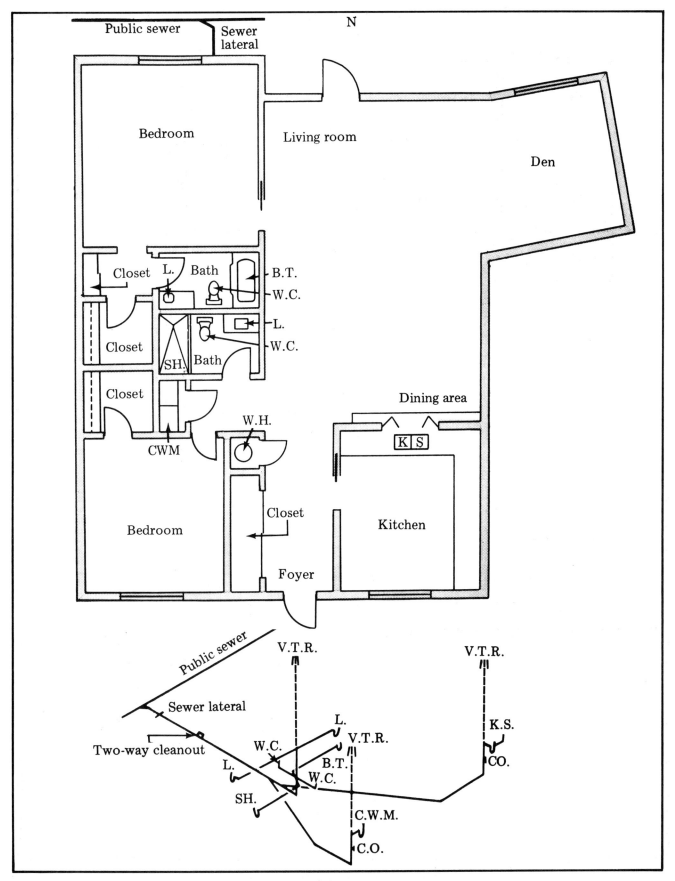

Single Family Residence Layout
Figure 2-12

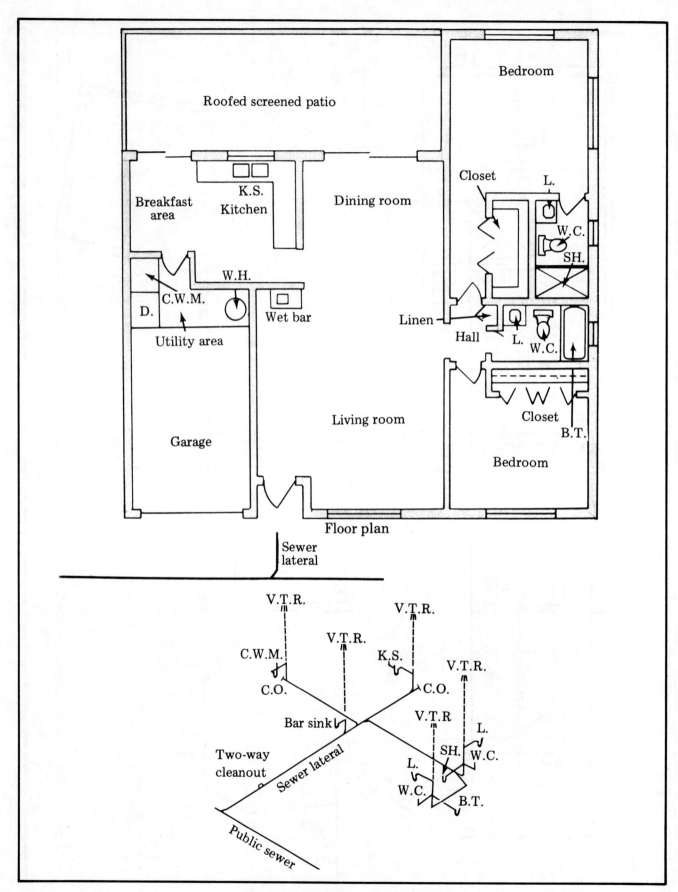

Single Family Residence Layout
Figure 2-13

Floor Plans & Layouts 25

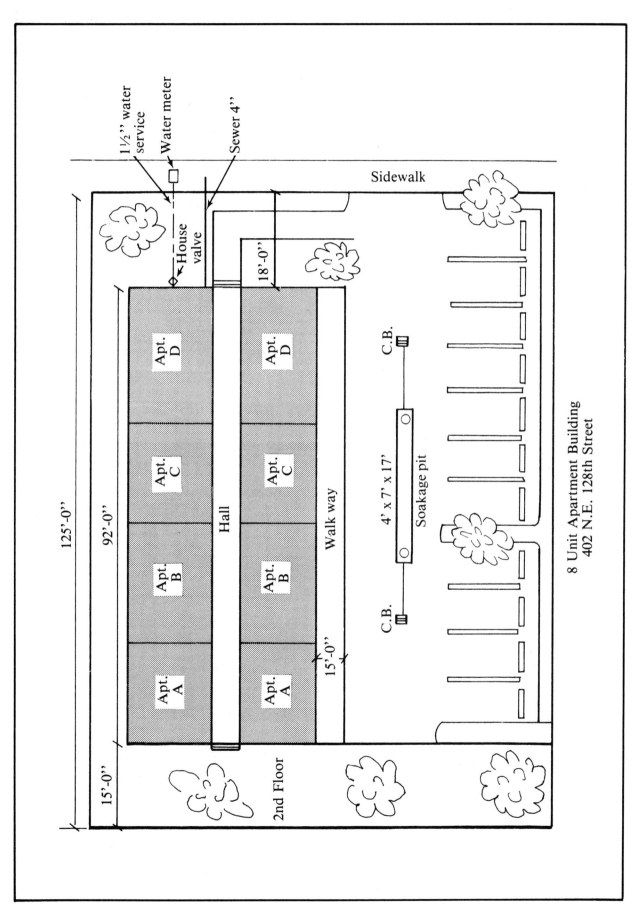

**Plot Plan
Figure 2-14**

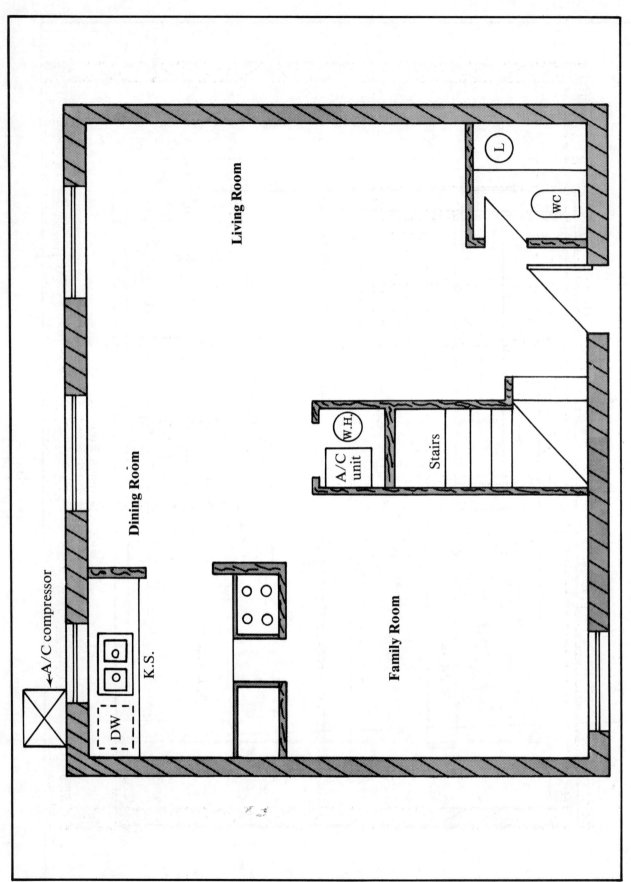

First Floor Apartment "C"
Figure 2-15

Floor Plans & Layouts

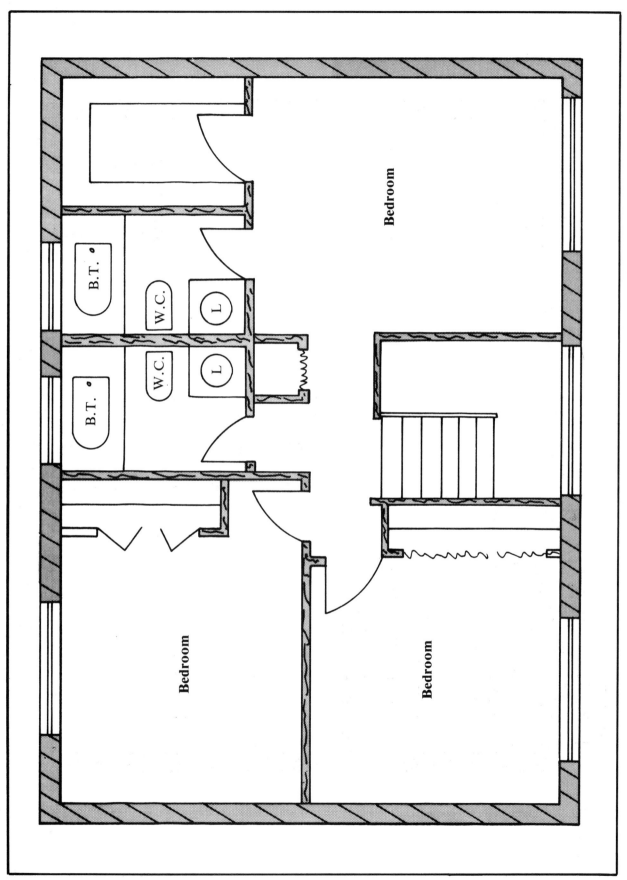

Second Floor Apartment "C"
Figure 2-16

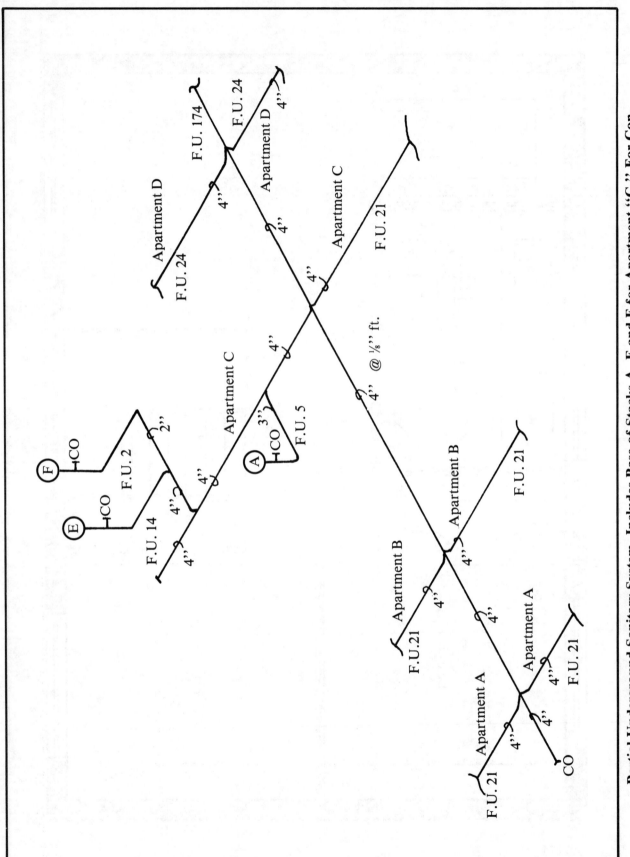

**Partial Underground Sanitary System. Includes Base of Stacks A, E and F for Apartment "C." For Continuation of Waste and Vent Stacks see Figure 2-18
Figure 2-17**

Floor Plans & Layouts

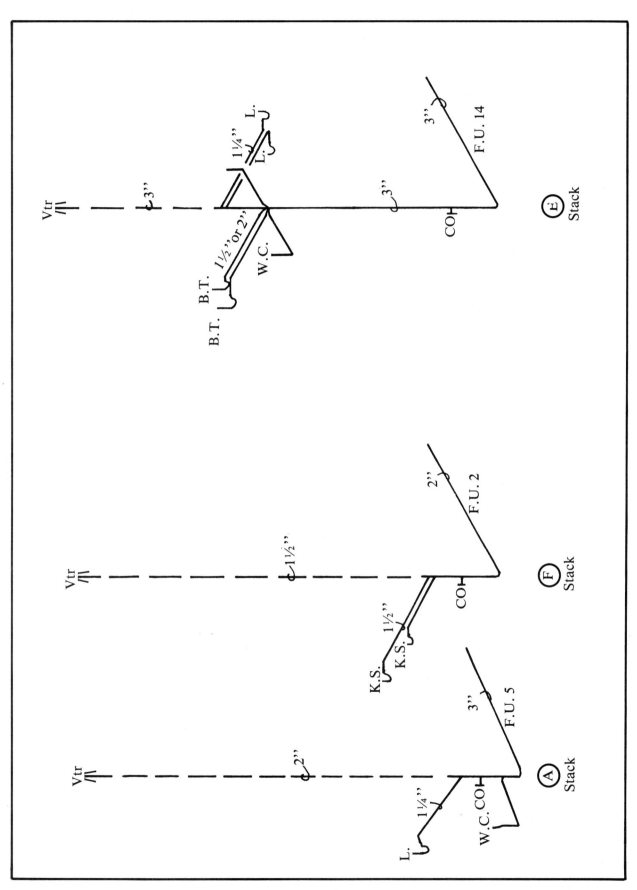

Waste and Vent Stacks A, E and F for Apartment "C"
Figure 2-18

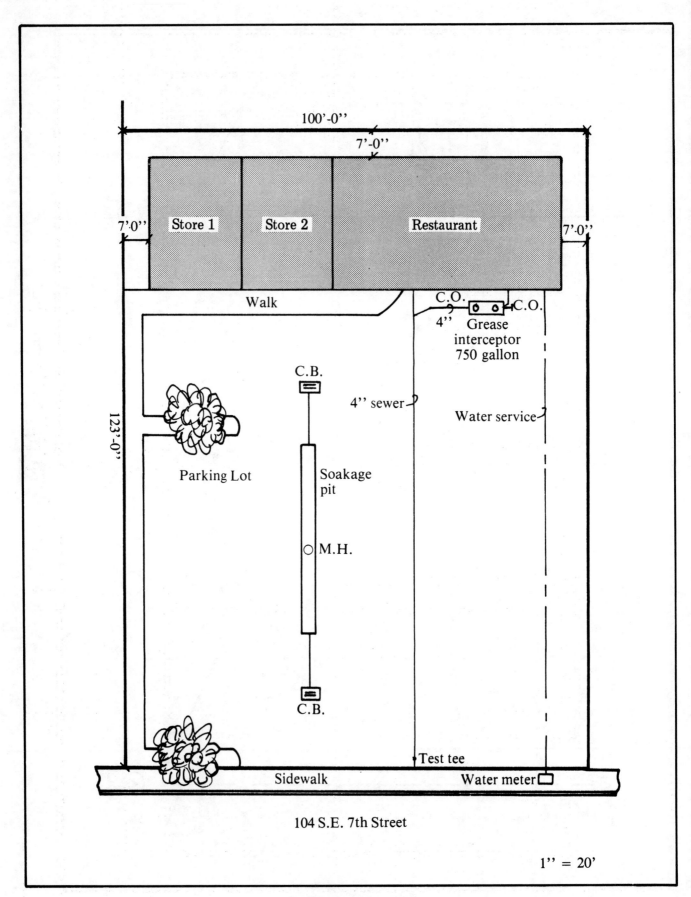

Commercial Plot Plan
Figure 2-19

Floor Plans & Layouts 31

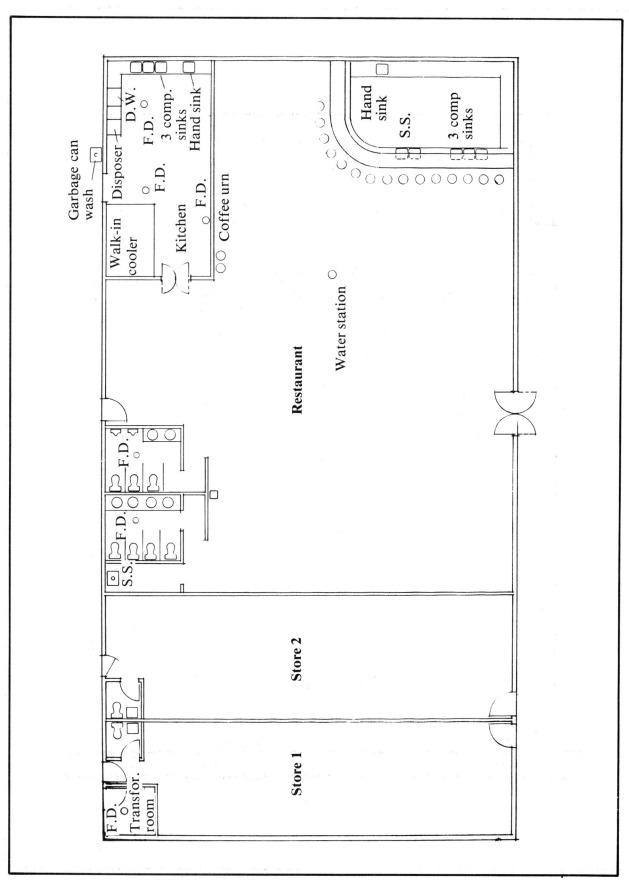

**Commercial Floor Plan
Figure 2-20**

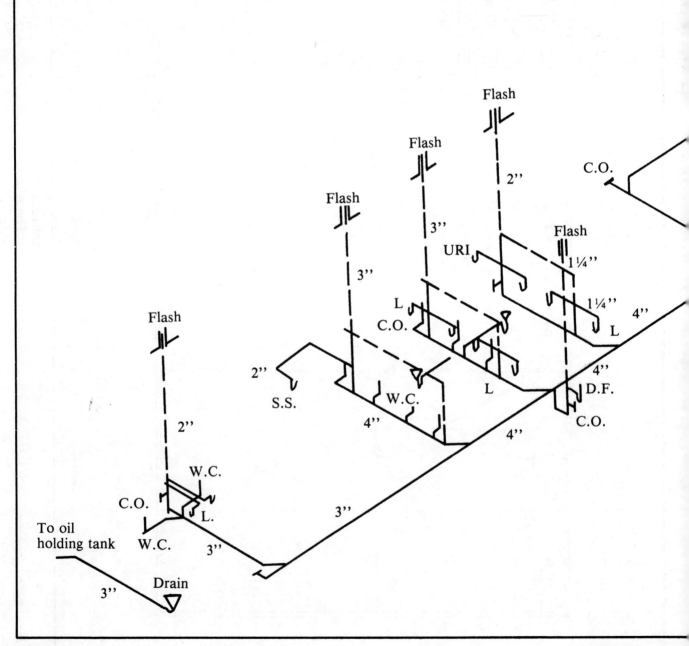

Sanitary Isometri

Floor Plans & Layouts

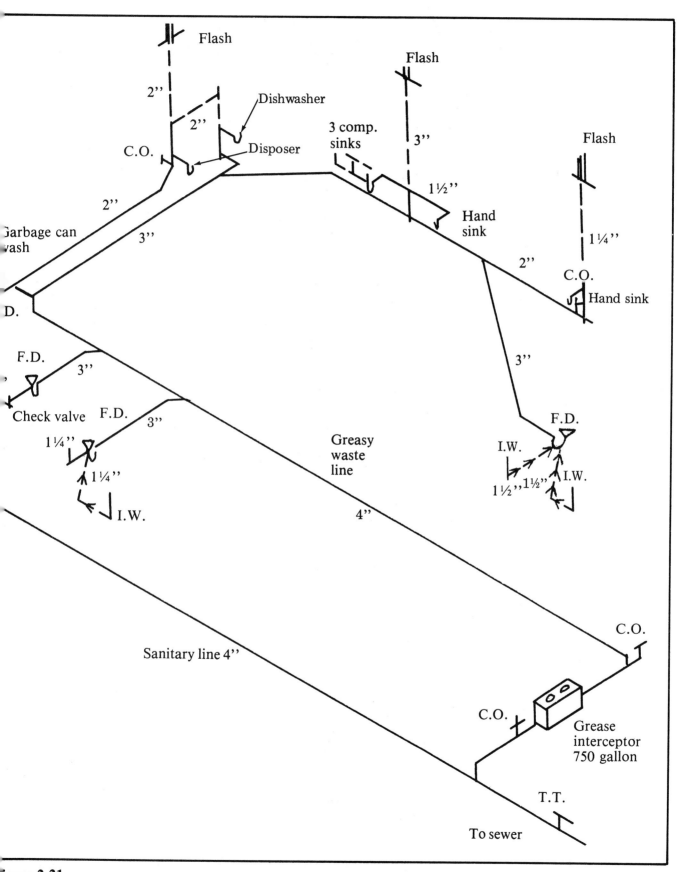

Figure 2-21

3

Isometric Drawings

Everyone who designs plumbing systems must know how to read and make isometric drawings. They are used by plumbing contractors to show piping layouts, indicate the number and type of fittings, and to compute the labor for DWV installations. They also show the plumber how to rough-in the plumbing system. Every plumbing estimator must have a good knowledge of what the applicable code requires for sanitary drainage and vent systems. You should be able to draw isometrics quickly and accurately that comply with the code. If you have trouble making isometric projections, study the following pages. Otherwise you may want to skip directly to Chapter 4. Reading and making isometric illustrations is a skill you can master quickly. Only three basic angles are needed to express the plumbing system: the horizontal pipe, the vertical pipe and the forty-five degree angle pipe. Figure 3-1 is a valuable guide to practicing these isometric angles. The only pieces of equipment necessary are a sharp No. 2 pencil and a 90-60 right triangle. By following the directions and practicing the exercises in this chapter, you will soon be able to produce your own isometric projections quickly and accurately, and be able to follow those made by others.

Making Isometric Drawings

First, draw a circle with a dot in the exact center. Place the letter N at the top of the page to designate north as shown in Figure 3-1. (Don't let the dotted line confuse you; it is used only to determine the proper angle necessary to illustrate the sanitary systems as shown in unbroken solid lines.) Using the 90-60 right triangle (see Figure 3-2), square the short base with the right edge of your paper and draw line A through the center dot. This represents A, the north-south horizontal pipe. Again using the 90-60 right triangle, square the short base with the left edge of your paper and draw line B through the center dot. This represents B, the east-west horizontal pipe. To draw C, the vertical line, square the short base of the 90-60 right triangle with the lower edge of the paper and use the long base as a straight edge. Connect line C with any of the horizontal lines as desired. To determine the placement of line E, bisect the area between the horizontal line B and the dividing dotted line. Then draw line E east and west through the center dot. (This line is necessary to show the change in direction assumed by the 45-degree fittings of either a wye or 1/8 bend.) Repeat the same procedure for line D north and south. The lower portion of Figure 3-1 shows a simple isometric drawing of all three basic angles used in designing rough plumbing for any building.

Fittings Within an Isometric Drawing

The lines on isometric drawings represent pipe and fittings. Symbols are used to show the types of fixtures that will be used. The symbols on the drawings are the same, regardless of the type of pipe and fittings used. Figures 3-3, 3-4, and 3-5 show typical isometric drawings. Each fitting in these drawings

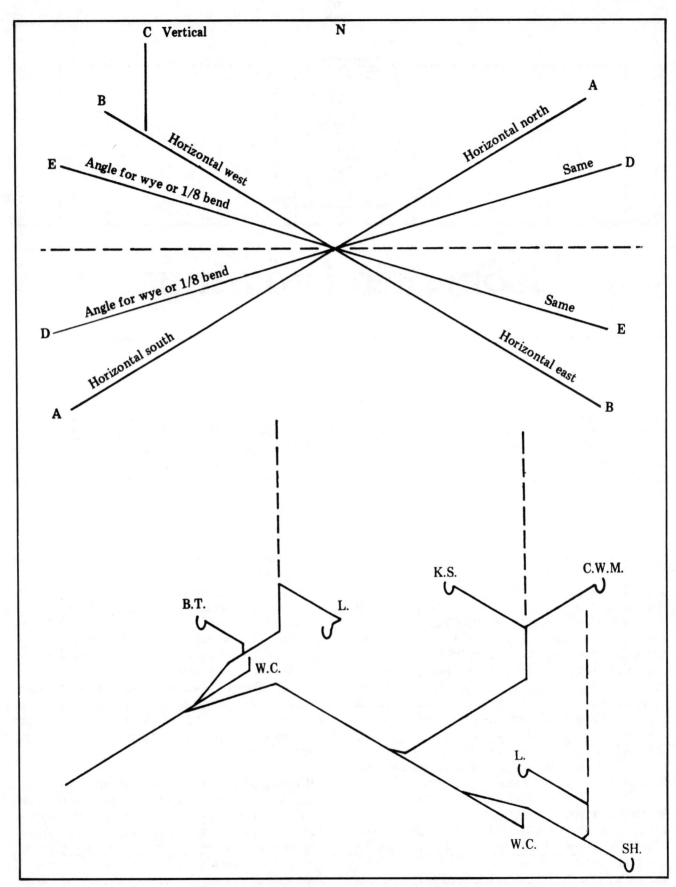

Isometric Drawing Illustrating the Various Angles
Figure 3-1

Isometric Drawings

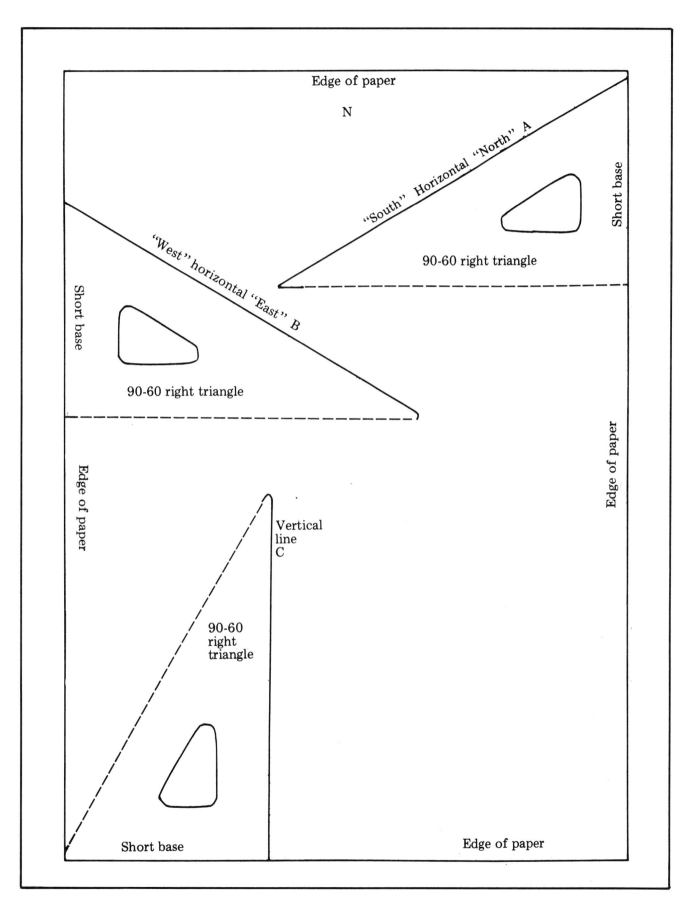

90-60 Right Triangle
Figure 3-2

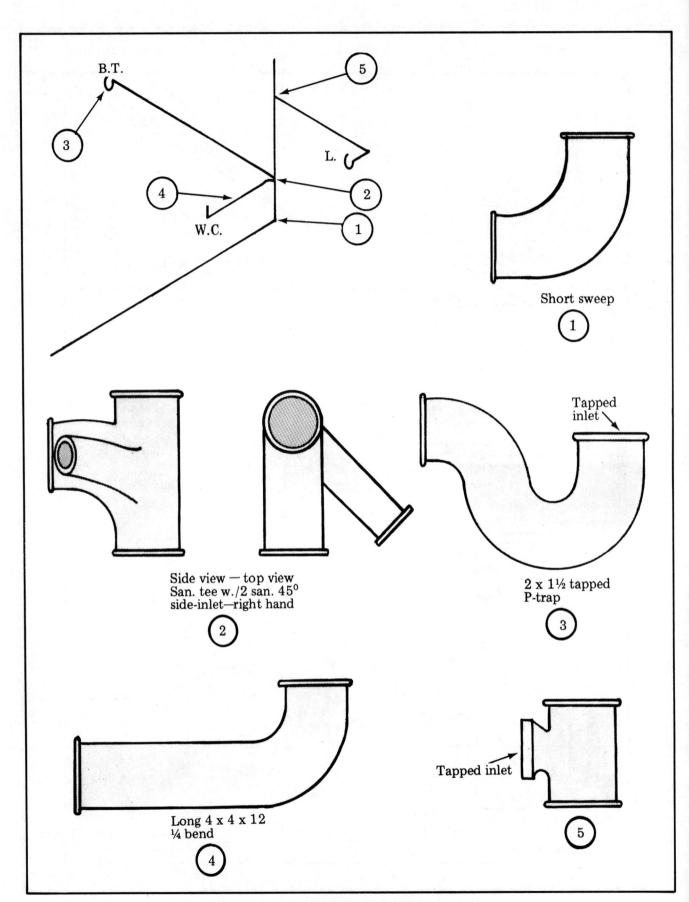

Typical Isometrics with Close-ups
Figure 3-3

Isometric Drawings

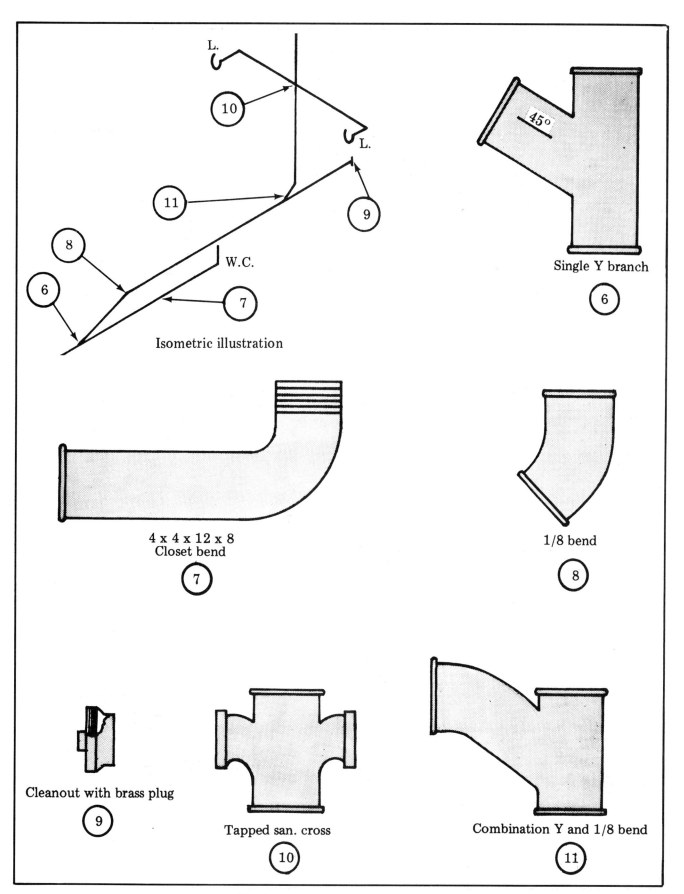

Typical Isometric with Close-ups
Figure 3-4

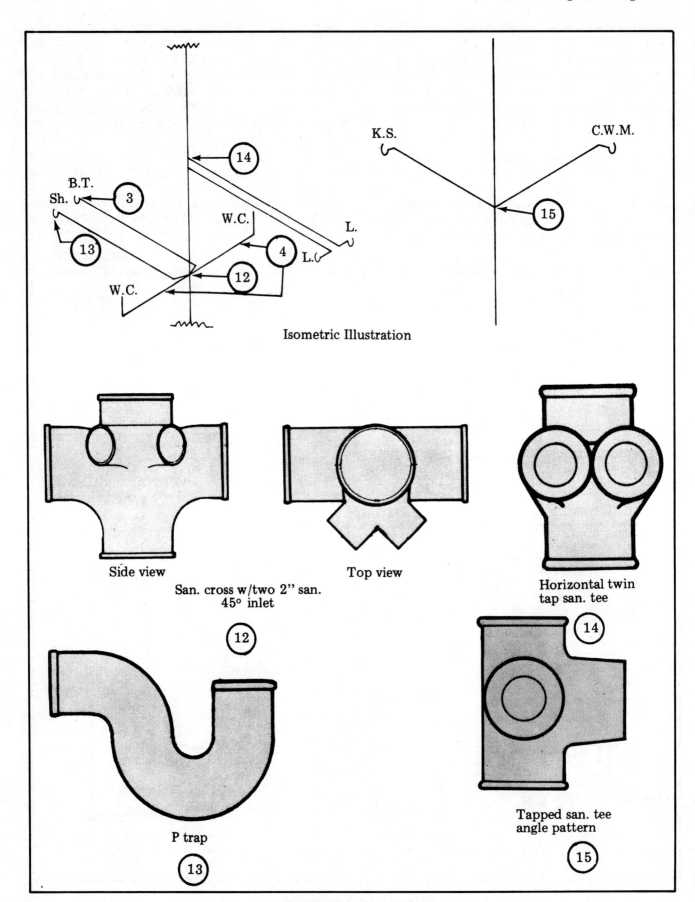

Typical Isometric with Close-ups
Figure 3-5

is numbered to correspond with the drawing below. Look at Figure 3-5. The horizontal twin tap sanitary tee (also known as an "owl fitting") is the same as fitting number 14 at the top of the page. This fitting connects two similar fixtures to the same waste and vent stack at the same level. In this case, it connects two lavatories.

Classes of Pipes and Fittings

The drainage system consists of three major classes of piping and fittings. The first class consists of all parts of the system necessary to receive and convey the used water to a public or private disposal system. The second class consists of the fixture traps. The trap is used for sanitary and health purposes. It is designed and constructed to provide a liquid seal to prevent odors, gases and vermin from entering at fixture locations. The third class consists of the vent pipes, which admit and emit air to and from all parts of the drainage system. Vent pipes are sized and arranged to relieve pressure that builds up as water is discharged into the system. The three classes of pipes and fittings are clearly illustrated in Figure 3-6, an isometric drawing of a simple sanitary system. The drainage system and the traps (the portion of the system that receives liquid waste) are illustrated by solid lines. The vents (the dry portions) are illustrated by broken lines.

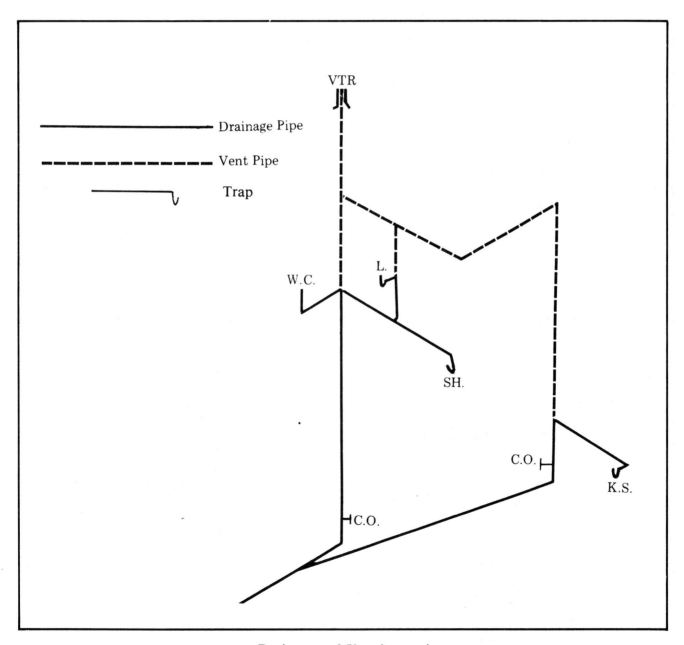

Drainage and Vent isometric
Figure 3-6

Part 2

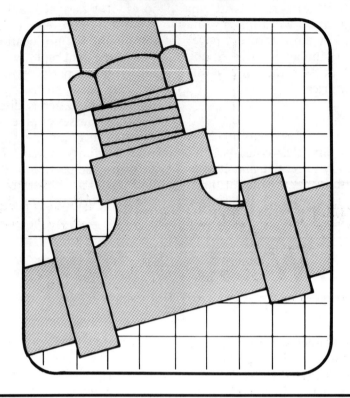

Special Plumbing Systems

- Interceptors & Special Waste Piping
- Trailer Park Systems
- Fire Protection Systems
- Gas Systems
- Sanitary Sitework & Water Distribution Systems

4

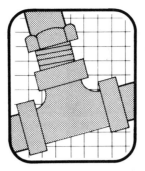

Interceptors & Special Waste Piping

Many substances are considered objectionable waste or harmful to the building drainage system, the public sewer, or the sewage-treatment plant. Such substances include grease, flammable waste, sand, plaster, lint, hair and ground glass. These materials must be intercepted or separated from the liquid waste before they enter the collection system.

Interceptors, special traps and special drain piping are an important part of plumbing work. This chapter helps the estimator accurately estimate these special plumbing systems. It should help prepare him for take-offs on most any type of special system job. All interceptors should be included in your estimate. This chapter will help you recognize when an interceptor is needed and understand the problems associated with installing each type of interceptor.

Interceptors

Many sizes and types of interceptors are available. Before installing interceptors on jobs not requiring approved plans (e.g. replacements), the estimator should prepare a detailed drawing and submit it to local authorities for advance approval.

When interceptors are required in drain lines, the estimator should remember that waste *not* requiring separation must *not* be discharged through an interceptor. There must be two distinct systems: one for special waste and one for sanitary waste. This is shown clearly in Figures 2-19, 2-20, and 2-21.

Nearly all interceptors required by modern codes are for commercial establishments. The three most often used interceptors are the *grease interceptor* (Figure 4-1), the *oil interceptor* (Figure 4-2), and the *lint interceptor* (Figures 4-3 and 4-4).

Grease Interceptors

Commercial buildings such as restaurants, hotel kitchens and bars, factory cafeterias and restaurants, clubs, processing plants and the like, must have a grease interceptor or trap to accumulate grease that might otherwise clog the waste lines. When required, a separate grease waste line is installed to serve pot, scullery, food scrap and vegetable preparation sinks. Floor drains that may receive kitchen waste spillage and floor drains that may receive waste from individual fixtures, appliances or other apparatus must go through an approved grease interceptor. However, waste from a commercial-type food grinder *must not pass through a grease interceptor*. It must pass directly into the main sanitary drainage system. (See Figure 2-21 in Chapter Two.)

The minimum capacity of grease interceptors for commercial establishments that prepare and serve food is generally determined by the seating capacity. (See Table 4-5.) If an establishment has a large seating capacity, it must have grease interceptors capable of handling large amounts of grease. These large capacity interceptors must be located outside the building they serve. Some codes allow floor mounted grease interceptors for commercial establishments that generate small amounts of grease. The capacity of these interceptors is rather

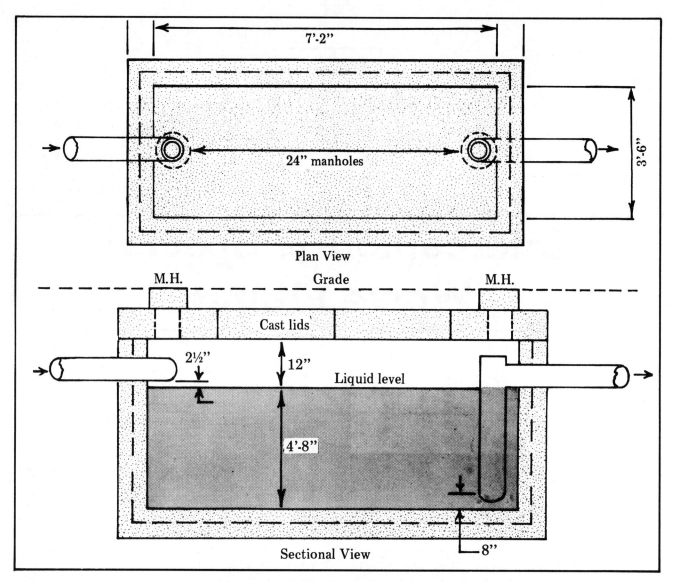

Grease Trap Detail — 750 — Gallons
Figure 4-1

small, generally 14 to 50 pounds. They can be located within the building and close to the fixtures they serve. Check your code for the permitted use. Figure 4-6 shows a floor-mounted grease interceptor.

Commercial establishments that prepare and sell food *only* on a take-out basis usually are not required to have a grease interceptor.

The materials from which grease interceptors may be constructed are governed by the local code. Floor-mounted interceptors generally have a cast iron body. Larger interceptors, located outside the building, are usually concrete. In some areas steel or fiberglass may also be used. The structural design criteria for grease interceptors are fairly uniform across the country. Figure 4-1 shows the principal design requirements.

Gasoline, Oil and Sand Interceptors

An interceptor must be provided in commercial buildings where the introduction of gasoline, grease, oil or sand into the drainage system is possible.

Bucket type floor drains should be used where this type of interceptor is required. These drains should have a minimum 4-inch diameter outlet. The bucket is removable for cleaning and is made of the same material as the floor drain. The bottom portion of the bucket is solid so it retains sand. Drainage holes near the top of the bucket let liquid waste pass out of the bucket and into the pipe or pipes leading to the interceptor.

Interceptors & Special Waste Piping

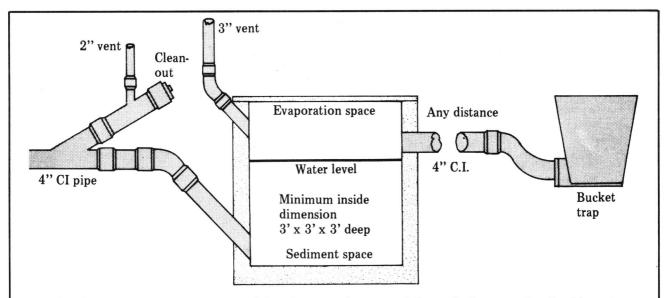

Interceptor for Gasoline, Oil, Sand, Auto Repair, Etc.
Figure 4-2

Oil interceptors usually must have a minimum capacity of 18 cubic feet per 20 gallons of design flow per minute. Figure 4-2 shows how the interceptor works and describes the conditions under which the venting may be omitted or required. Oil interceptors may be constructed of concrete at the job site or purchased as a preassembled unit of steel or other approved material.

Slaughter House Interceptors
Where an establishment slaughters, prepares or processes meat, the waste from the floors must pass through a specially designed floor drain before entering the grease interceptor. These bucket type floor drains are equipped with metal screen baskets that keep solids more than 1/2 inch across out of grease interceptor traps. Fats and smaller solids are collected in the interceptor trap itself. The structural design criteria are the same as for regular grease interceptors.

Laundry Interceptors
Commercial laundries discharge solids such as lint, string, and buttons with the liquid waste. Solids such as these should not enter the drainage system. Therefore, lint interceptors must be installed on laundry drainage pipes.

The lint interceptor should have a *nonremovable* 1/2-inch mesh screen, metal basket or a similar device to collect the solids. The screen should be constructed to allow easy cleaning. (See Figure 4-4.)

The plumbing code considers the horizontal drainage pipes serving commercial clothes washing machines to be indirect waste pipes. This is a unique method of piping but is economical and practical in this application. The indirect waste system does not have to be trapped or vented as do most other plumbing fixtures. The washing machine standpipes are open-ended 3- or 4-inch diameter pipes and extend to about 26 inches above the

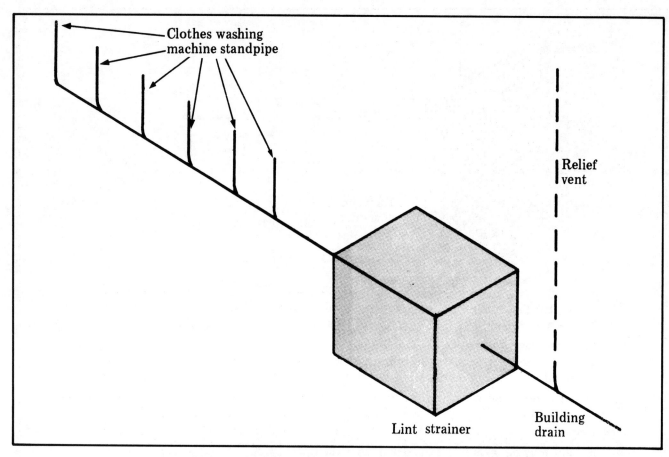

Install Lint Interceptors on the Drainage Pipes From Laundries
Figure 4-3

finished floor. These drain pipes receive the discharge from the washers through flexible hose. A 3-inch standpipe can accommodate two machines; a four inch standpipe will serve four machines.

Horizontal drain pipes collect waste from the standpipes and convey it to a lint interceptor which is generally one pipe above the liquid level. The drainage system is not vented because standpipes permit free circulation of air. (See Figure 4-3.)

A lint interceptor is usually constructed in place on the job site by the plumbing contractor. Figure 4-4 should help you understand its design.

The outlet or discharge pipe extends down at a 45-degree angle to approximately two inches from the bottom of the lint interceptor. This serves two purposes: first, light objects that have passed through the screen and are floating in the top portion of the interceptor water will not sink into the outlet pipe and thus into the building's sanitary drainage system. Second, the extension of the outlet pipe toward the bottom of the lint interceptor creates a liquid seal. This prevents sewer gases from entering the building through the clothes washing machine standpipes. The screen or screens retain foreign objects that can not pass through them. These objects should be removed regularly when the system is in service.

The outlet pipe from the lint interceptor is connected to the regular sanitary drainage system serving other fixtures within the building. A vent must be installed as close as possible to the lint interceptor on the horizontal discharge (outlet) pipes. (See Figure 4-4.) The vent pipe serves the drainage piping between the lint interceptor and the main building sewer or drainage system, supplying and removing air as needed. This keeps the lint interceptor from becoming air locked and ensures a free flow of waste water.

Other Trap Requirements

Because of the large quantity of broken glass or other solids generated in bottling plants, they are required to discharge their process wastes through an interceptor trap. These interceptors are designed to separate broken glass and other solids from liquid wastes.

Where large quantities of hair may be introduced

Interceptors & Special Waste Piping

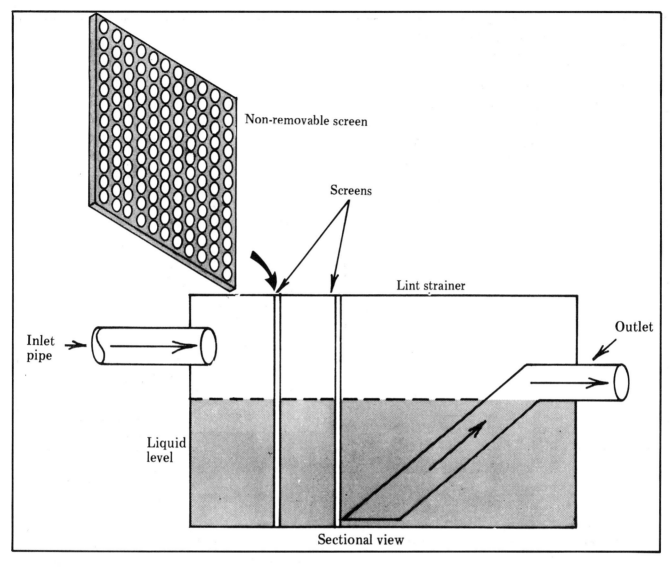

Lint Interceptor
Figure 4-4

Liquid Capacity of Grease Interceptor in Gallons			
Up to 25 Persons	25 to 50 Persons	51-100 Persons	101-150 Persons
600	750	1,200	1,600

Minimum Liquid Capacity of Grease Interceptor For Eat and Drink Establishments
Table 4-5

into the drainage system, interceptors similar to ones used for commercial laundries or swimming pools are required. Animal hospitals and dog grooming establishments, for example, may be required to install hair strainers (interceptors) on waste lines from bathtubs or other receptacles where animals are bathed.

In dental and orthopedic sinks where plaster, wax or other objectionable substances will be discharged into a drainage system, an interceptor trap must be installed in the waste line. (See Figure 4-7.)

Dilution or Neutralizing Tanks
Sanitary drainage systems are not generally constructed of materials that can withstand the corrosive action from many chemicals, acids, or strong alkalis. Corrosive liquids must pass through an approved dilution or neutralizing tank before discharge into the regular sanitary system. The neutralizing tank is usually constructed of glass, earthenware, or other non-corrosive materials. The

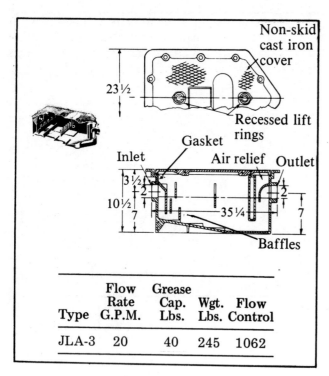

Grease Interceptors
Figure 4-6

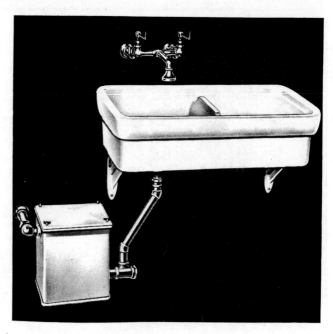

Plaster Work Sink—Integral Right Hand Tray—Vitreous China
Figure 4-7

tank should have a controlled water supply so that the harmful material can be diluted to the point where it will not harm the plumbing system.

The pipes, fittings, and vent system for corrosive waste must be installed independent of any other drainage or vent system and must be of duriron, plastic or other approved non-corrosive materials.

After being properly neutralized, the corrosive waste may be discharged into the regular sanitary drainage system.

Special Waste Piping
Indirect Waste Piping

Fixtures, appliances and devices not regularly classed as plumbing fixtures may be drained by indirect means if they have drips or drainage outlets. These types of installations include refrigerators, ice boxes, bar sinks, cooling or refrigerating coils, laundry washers (commercial), extractors, steam tables, egg boilers, coffee urns, stills, sterilizers, water stations, water lifts, expansion tanks, cooling jackets, drip or overflow pans, air conditioning condensate drains, drains from overflows, and relief vents from the water supply system.

Indirect drainage prevents sewage from backing up into these special fixtures and contaminating the contents if there is a stoppage in the sanitary drainage system. For example, overflow and relief pipes on water supply systems, relief pipes on the water supply system and relief pipes on expansion tanks, sprinkler systems, and cooling jackets must always be indirectly connected to the sanitary drainage system. This avoids the possibility of a cross connection which would contaminate the potable water supply system. A positive separation by indirect means is required between the waste outlet and drainage inlet of hospital equipment and food storage and preparation establishments. This avoids contamination. This method of drainage is unique and very practical for fixtures with low discharge rates.

Indirect waste piping must be sized and installed to accommodate the outlet drainage of the fixture or appliance it serves. For example, a coffee urn is a special fixture which comes from the manufacturer equipped with a 1/2- or 3/4-inch drain from the drip pan. It has a waste drain smaller than 1¼ inches and thus would be rated as 1 fixture unit.

Refer to special fixture Table 4-8. List the number and types of fixtures to be used in the job you are estimating. The table shows fixture unit values so you can properly size all indirect waste piping for any building.

Indirect waste piping without its own trap need not be vented but must *never* connect directly to the sanitary drainage system. Indirect waste piping is designed to carry waste to a receiving fixture such as a floor sink or floor drain which does not receive material with highly offensive odors. Accessible cleanouts should be provided to allow cleaning and flushing.

Interceptors & Special Waste Piping

Fixture Drain or Trap Size in Inches	Fixture Units Value
1¼ and smaller	1 F.U.
1½	2 F.U.
2	3 F.U.

Special Fixtures
Table 4-8

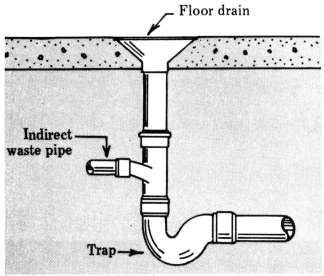

Below Floor Indirect Waste Pipe
Figure 4-9

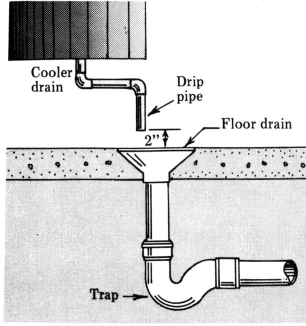

Above Floor Indirect Waste Pipe
Figure 4-10

Indirect waste pipe should be installed below the floor, where practical. If an indirect waste pipe is installed, it should connect through the receiving fixture above the water seal of the trap. See Figure 4-9.

When above-the-floor installation is required, the outlet should terminate one to two inches above the receiving fixture as in Figure 4-10.

Indirect waste pipe installed above the floor should have at least a 3/4 inch diameter. But it must never be smaller than the outlet drains of fixtures or appliances it serves. When installed below the floor, indirect waste pipe must be at least 1¼ inches in diameter.

Floor drainage from a walk-in refrigerator or food storeroom floor should be equipped with a flap check valve (see Figure 4-11) to protect these areas from contamination if there should be a sewage back-up.

The drip pipe from a walk-in refrigerator or a cooler room refrigeration coil should terminate one to two inches above the receiving fixture. A flap check valve in this case is not required.

Air Conditioning Condensate Drains

The minimum size indirect waste piping for air conditioning units below a floor slab is 1¼ inches. All risers passing through the slab must be sleeved. Some codes require that condensate drainage systems be vented.

Air conditioning units up to 5-ton capacity can discharge their waste on a pervious area such as bare soil. Units over 5 tons but under 10 tons may discharge their waste into a buried 10-inch diameter by 24-inch long pipe filled with 3/4-inch rock. No cover is required.

The main condensate drain line from a 10-ton or larger unit must discharge into a drainage well, storm sewer, adequate size soakage pit, drainfield or the building drainage system. The condensate drain from the air conditioning equipment *can not* discharge its waste on the roof of a building or into rain leaders that terminate on a sidewalk at grade level.

Storm Drainage System

Although professional engineers are usually responsible for sizing storm drainage systems for large buildings, you may have to size a smaller storm drainage system for your estimate. Therefore, you should know what principles are involved.

The sizing of all storm water drainage piping is based on the square footage of the impervious area (roof, parking lot, etc.) being served. A second im-

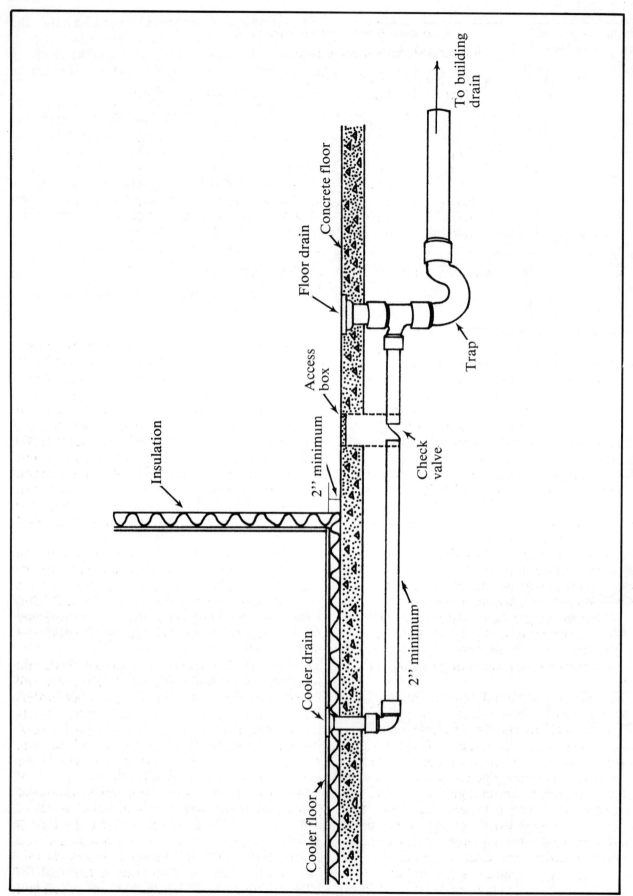

**Special Cooler Floor Drainage
Figure 4-11**

Nominal Pipe Size (Inches)	Maximum Roof Area (Square Feet) Building Storm Sewers & Drains			Gutters	Leaders
	⅛" per Ft. Slope	¼" per Ft. Slope	½" per Ft. Slope		
1½	127	190	222	---	222
2	270	380	460	---	460
2½	413	610	700	---	700
3	747	1,080	1,270	635	1,270
4	1,560	2,210	3,080	1,540	3,080
5	2,810	4,000	5,620	2,810	5,620
6	4,450	6,290	8,880	4,440	8,880
8	9,460	13,760	18,950	9,975	18,950

Size of Stormwater Drain, Leaders, and Gutters
Table 4-12

portant consideration is the maximum anticipated rainfall rate in any hour.

Code requirements vary because of fluctuation in maximum anticipated rainfall in any hour. Rainfall estimates are as little as two inches in some areas and more than eight inches in other areas. It follows, then, that the less rainfall, the smaller the pipe required; the greater the rainfall, the larger the pipe required.

Most tables in code books for sizing storm water drains, leaders, and gutters are already calculated for the geographic area involved. If you need to calculate storm water drains, leaders, or gutter sizes, use the following procedure.

Table 4-14 at the end of this chapter gives the maximum rates of rainfall for various cities. Codes are based on the maximum rate of rainfall of four inches per hour. *When maximum rates are more or less than four inches per hour, the figures for the area to be drained are adjusted by multiplying the square footage of drainage area by four and dividing the answer by the maximum rainfall in the local area.*

To get a better understanding of the procedure required for sizing storm water drainage pipes, refer to Table 4-12, from a model plumbing code, and Figure 4-13. The first column of Table 4-12 shows the pipe size of the horizontal building storm drain or the vertical leader pipe to the roof. The second, third, and fourth columns give the maximum square feet of roof area permitted, according to slope, for the pipe sizes listed in the first column. The sixth column gives the maximum square feet of roof area permitted for the leader size listed in column one. The fifth column is for gutters and downspouts that may or may not be in your bid price.

As an example, consider a roof with 900 square feet. This would be a building 20 feet wide by 45 feet long. Let's assume one leader is required and 1/8-inch slope per foot is the maximum pitch permitted. We now have the basic facts necessary to size the building storm drain and the leader pipe. Checking column two in the table, you see that 900 exceeds 747, the maximum square footage of roof area allowed on a 3-inch horizontal storm drain. Yet it does not exceed the 1,560 square feet allowed on a 4-inch horizontal storm drain. Thus, the horizontal storm drain for this building would be 4 inches.

To size the leader we use column six. Again, we see that 900 exceeds 700, the maximum square feet of a roof area allowed on a 2½-inch leader. Nine hundred square feet does not exceed the 1,270 square feet permitted on a 3-inch vertical leader pipe. Therefore, the vertical leader pipe would be 3 inches.

Figure 4-13 shows a vertical wall that sheds rain water on a main building roof. The vertical wall section must be considered in sizing the rain leaders and the horizontal storm water pipes. Many building codes require that 1/3 of the total area of the vertical wall be added to the area of the main roof. Other codes require that more or less of the vertical wall section be included.

Refer to the illustration in the upper portion of Figure 4-13. Rain water must be disposed of from a roof area of 5,300 square feet. Since the roof is equally divided and has two roof drains and two leaders, each leader must be sized to carry the rain water from an area of 2,650 square feet (half of 5,300 square feet). In Table 4-12, sixth column,

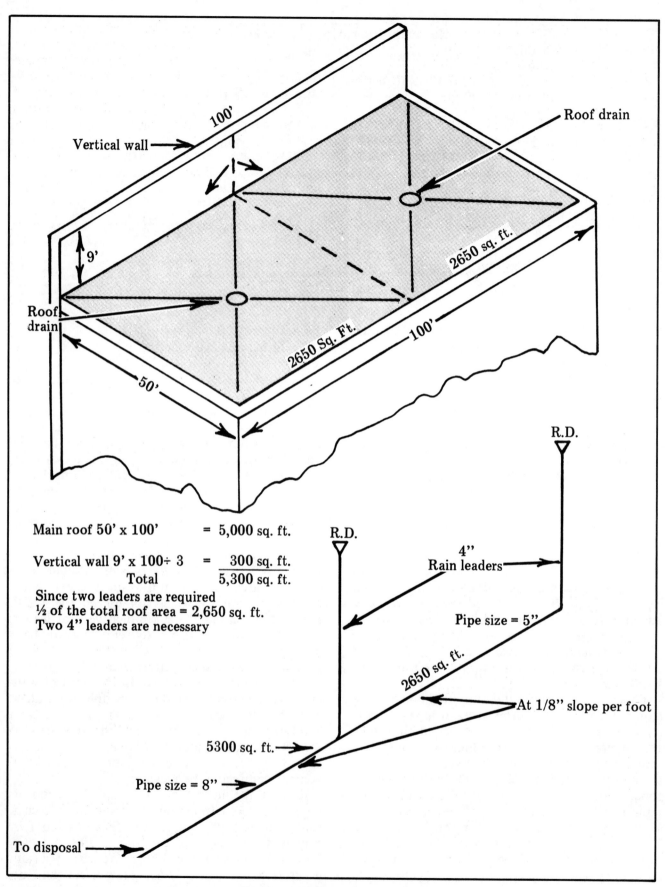

Stormwater Drainage System
Figure 4-13

Interceptors & Special Waste Piping

you see that the next highest square foot area is 3,080. The first column shows that a 4-inch leader would be required to serve each roof drain.

The horizontal storm water pipes to serve these leaders at 1/8 inch slope per foot are sized from column two. Since the end section will convey rain water from a single leader of 2,650 square feet, the figure 2,810 square feet (column two) and the corresponding 5-inch diameter in column one must be used. Therefore, this section of horizontal pipe would be 5 inches.

The second section of horizontal storm water pipes will convey water from the complete roof area of 5,300 square feet. Therefore, the figure 9,460 square feet (column two) and the corresponding 8-inch diameter in column one must be used. The horizontal section of piping that conveys the storm water load from the entire roof of this building to the point of disposal must be 8 inches.

The plans may show the size of storm drains. But if they don't, you must calculate the correct size and include the cost of that drain in your estimate.

Maximum Rates of Rainfall for Various Cities, in inches per hour

City	Rate	City	Rate	City	Rate	City	Rate
Alabama:		**District of Columbia:**		Des Moines	6.4	**Minnesota:**	
Anniston	7.2	Washington	7.2	Dubuque	7.4	Duluth	6.2
Birmingham	7.0	**Florida:**		Keokuk	6.8	Minneapolis	6.6
Mobile	8.4	Apalachicola	7.3	Sioux City	7.0	Moorhead	5.8
Montgomery	7.0	Jacksonville	7.4	**Kansas:**		St. Paul	6.3
Alaska:		Key West	6.6	Concordia	7.5	**Mississippi:**	
Fairbanks	3.7	Miami	7.5	Dodge City	6.3	Meridian	7.4
Juneau	1.7	Pensacola	9.4	Iola	8.4	Vicksburg	7.5
Arizona:		Sand Key	6.6	Topeka	6.8	**Missouri:**	
Phoenix	4.3	Tampa	8.4	Wichita	6.9	Columbia	7.0
Arkansas:		**Georgia:**		**Kentucky:**		Hannibal	6.5
Bentonville	7.4	Atlanta	7.7	Lexington	6.0	Kansas City	6.9
Ft. Smith	6.2	Augusta	8.4	Louisville	7.0	St. Joseph	6.5
Little Rock	6.7	Macon	7.2	**Louisiana:**		St. Louis	6.5
California:		Savannah	6.8	New Orleans	8.2	Springfield	7.0
Eureka	2.7	Thomasville	7.3	Shreveport	7.5	**Montana:**	
Fresno	3.6	**Hawaii:**		**Maine:**		Havre	4.3
Los Angeles	3.6	Honolulu	5.2	Eastport	4.7	Helena	3.8
Mt. Tamalpais	2.5	**Idaho:**		Portland	4.7	Kalispell	3.3
Pt. Reyes	2.4	Boise	2.7	**Maryland:**		Miles City	7.0
Red Bluff	3.8	Lewiston	3.1	Baltimore	7.8	Missoula	2.7
Sacramento	3.0	Pocatello	3.7	**Massachusetts:**		**Nebraska:**	
San Diego	3.3	**Illinois:**		Boston	5.5	Lincoln	6.6
San Francisco	3.0	Cairo	6.6	Nantucket	4.8	North Platte	6.0
San Jose	2.0	Chicago	7.0	**Michigan:**		Omaha	7.0
San Luis Obispo	3.1	Peoria	6.2	Alpena	6.1	Valentine	6.3
Colorado:		Springfield	6.6	Detroit	6.4	**Nevada:**	
Denver	5.7	**Indiana:**		East Lansing	6.1	Reno	3.2
Grand Junction	3.0	Evansville	6.0	Escanaba	5.4	Tonopah	3.0
Pueblo	5.0	Ft. Wayne	6.3	Grand Haven	5.0	Winnemucca	2.7
Wagon Wheel Gap	3.6	Indianapolis	6.3	Grand Rapids	6.0	**New Hampshire:**	
		Terre Haute	7.5	Houghton	5.0	Concord	6.2
Connecticut:		**Iowa:**		Marquette	6.0	**New Jersey:**	
Hartford	6.2	Charles City	6.5	Port Huron	5.3	Atlantic City	6.2
New Haven	6.6	Davenport	6.4	Sault Ste. Marie	5.2		

Table 4-14

Sandy Hook	7.0	Cleveland	6.9	Pierre	6.5	Northfield	6.2
Trenton	6.4	Columbus	6.1	Rapid City	5.5	Virginia:	
New Mexico:		Dayton	6.0	Yankton	5.8	Cape Henry	7.4
		Sandusky	6.2	Tennessee:		Lynchburg	6.0
Albuquerque	3.7	Toledo	6.0	Chattanooga	7.2	Norfolk	6.8
Roswell	5.4	Oklahoma:		Knoxville	6.2	Richmond	7.2
Santa Fe	4.4	Oklahoma City	6.7	Memphis	6.8	Wytheville	5.7
New York:		Oregon:		Nashville	7.2	Washington:	
Albany	6.0	Baker	3.3	Texas:		North Head	2.8
Binghamton	5.0	Portland	3.0	Abilene	7.2	Port Angeles	2.2
Buffalo	5.5	Roseburg	3.6	Amarillo	6.8	Seattle	2.2
Canton	5.6			Austin	7.4	Spokane	3.1
Ithaca	6.0	Pennsylvania:		Brownsville	7.5	Tacoma	2.8
New York	6.6	Erie	6.5	Corpus Christi	6.6	Tatoosh Island	3.2
Oswego	5.9	Harrisburg	7.0	Dallas	7.2	Walla Walla	2.7
Rochester	5.4	Philadelphia	6.5	Del Rio	7.6	Yakima	2.6
Syracuse	6.3	Pittsburgh	6.4	El Paso	4.2	West Virginia:	
North Carolina:		Reading	6.5	Fort Worth	6.6	Elkins	6.2
Asheville	6.7	Scranton	6.1	Galveston	8.2	Parkersburg	6.7
Charlotte	7.0	Puerto Rico:		Houston	8.0	Wisconsin:	
Greensboro	6.6	San Juan	5.7	Palestine	6.3	Green Bay	5.1
Hatteras	6.8	Rhode Island:		Port Arthur	7.5	LaCrosse	6.5
Raleigh	7.5	Block Island	5.3	San Antonio	7.5	LaCrosse	6.5
Wilmington	7.0	Providence	4.8	Taylor	7.7	Madison	6.0
North Dakota:		South Carolina:				Milwaukee	6.2
Bismarck	6.7	Charleston	7.0	Utah:		Wyoming:	
Devils Lake	6.8	Columbia	6.6	Modena	3.8	Cheyenne	5.6
Williston	6.5	Greenville	6.6	Salt Lake City	3.4	Lander	3.7
Ohio:		South Dakota:		Vermont:		Sheridan	5.2
Cincinnati	6.5	Huron	6.2	Burlington	5.4	Yellowstone Park	2.5

Rates given are intensities for a 5 minute duration and a 10 year return period, from Technical Paper Number 25, Rainfall Intensity-Duration-Frequency Curves, U. S. Dept. of Commerce, Weather Bureau.

Taken directly from *National Standard Plumbing Code*, 1973. Published by National Association of Plumbing-Heating-Cooling Contractors. Co-sponsored by National Association of Plumbing-Heating-Cooling Contractors, 1016 20th Street, N.W., Washington, D.C. 20036 and American Society of Plumbing Engineers, 16161 Ventura Blvd., Suite 105, Encino, California 91316.

Table 4-14 (continued)

5

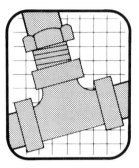

Trailer Park Systems

The popularity of both the mobile home and the travel trailer or motor home has forced local code adopting bodies to compile codes which apply to trailer parks. Many, if not most, codes now establish minimum sanitary plumbing facilities and installation methods specifically for trailer parks. These standards vary considerably from requirements for conventional permanent structures. The plumbing estimator should understand the standards established for designing and installing sanitary collection and water distribution systems in trailer parks.

Sooner or later most plumbing estimators will have to estimate this type of work. The information and illustrations in this chapter will help you understand and meet most model code requirements.

Trailer Park Sanitary Facilities
Sites approved for either independent or dependent trailers must have a service building with toilet facilities within 200 feet of the most distant trailer site. The code definitions will help you understand the differences between the two types of trailers and their sanitary facilities.

A *dependent travel trailer* is any trailer coach used as a temporary dwelling unit for travel, vacation and recreation. It usually has built-in sanitary facilities but not a plumbing system suitable for connection to park sewage and water supply systems. It has a kitchen, bath, and living quarters and uses a water storage tank to operate the plumbing fixtures and a holding tank to retain the waste water.

An *independent trailer coach* is any trailer coach designed for permanent occupancy that has kitchen and bathroom facilities and a plumbing system suitable for connection to the park sewage and water supply system. The operator of the trailer park provides each trailer coach space with water, electricity, and a gas and water-tight connection for sewage disposal. See Figures 5-3 and 5-4 for different types of independent trailer coach sewer connections.

Trailer parks specializing in services for dependent trailers must provide a service building or buildings with a minimum number of plumbing fixtures for both sexes as shown in Table 5-1.

Parks completely sewered and intended to service both dependent and independent trailers must provide a service building or buildings with a minimum number of fixtures for each 100 trailer spaces or fraction thereof, as shown in Table 5-2.

All parks must provide laundry facilities as follows: For each 25 trailer coach spaces, an automatic washer and one 2-compartment laundry tray. Where wringer-type washing machines are used, one 2-compartment laundry tray must be provided for each machine. For example, if a park has 100 trailers, it must provide four automatic washers and one 2-compartment laundry tray, *or* four wringer-type washers and four 2-compartment laundry trays.

In determining the facilities required, the

Item	Women	Men
1 water closet for each	15	20
1 lavatory for each	20	20
1 shower bath for each	20	20
1 urinal for each	---	25

Table 5-1

For Women	For Men
1 water closet	1 water closet
1 lavatory	1 lavatory
1 shower	1 shower
---	1 urinal

Table 5-2

estimator must assume that each trailer will be occupied by three people.

Sizing a Park Drainage System

Fixture unit load value, which must be assumed for each trailer, can be as low as 6 or as high as 15, depending on the code being used. Typically, codes use a value of 9 fixture units per trailer. This procedure differs considerably from conventional buildings where each plumbing fixture is totaled separately. Because the number of fixture units assumed is only an estimate, most codes include a table for sizing the park drainage system based on the number and type of trailers. Table 5-5 is typical of what many codes require. It shows the maximum number of trailers that may be connected to each sewer size.

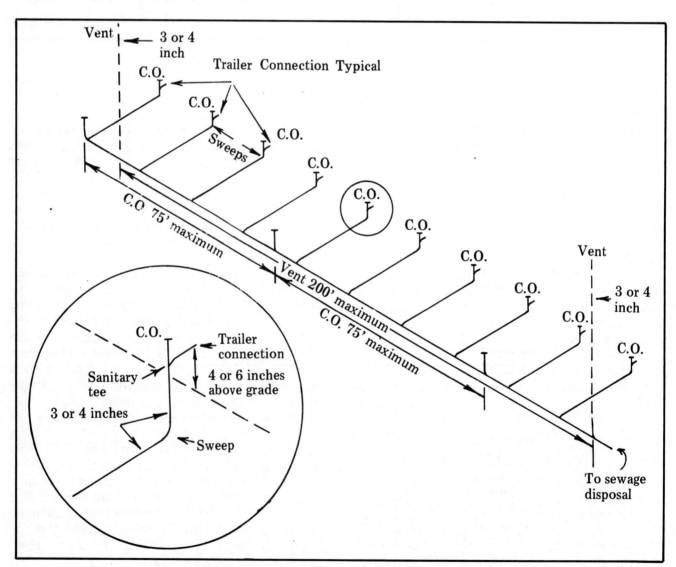

Sewage Collection System for Properly Trapped and Vented Trailers
Figure 5-3

Trailer Park Systems

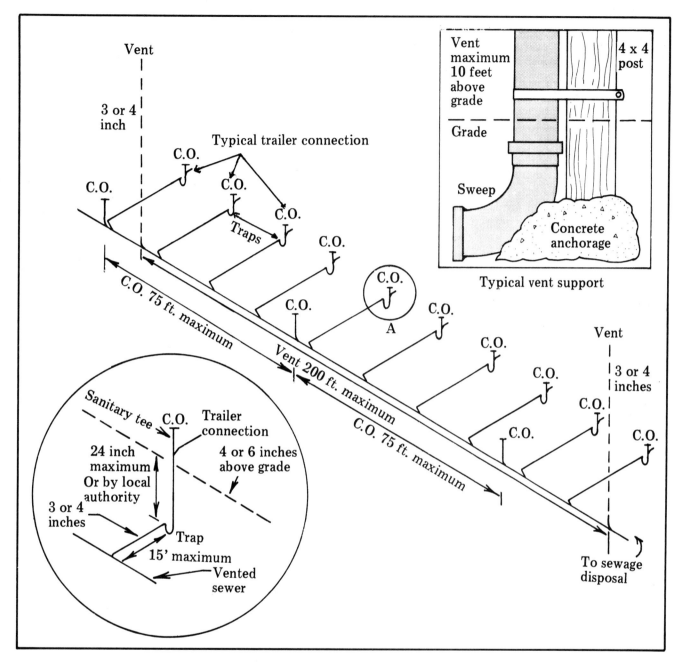

Sewage Collection System for Trailers Not Properly Trapped and Vented
Figure 5-4

Estimating Trailer Parks Systems

Estimate bedding and backfill costs the same as building sewers. Costs will vary with the type of materials used. A sewer lateral not less than 3 inches in diameter (4 inches in some codes) must be provided for each trailer site. The line must be capped when not in use. Vent pipes on building drainage systems (see Figures 5-3 and 5-4) are located at least ten feet from adjoining property lines and must extend ten feet above ground level. Vent pipes should be securely anchored to the equivalent of a 4 x 4 post (see Figure 5-4) driven into the ground. Supports should be resistant to rot and deterioration.

The first vent should be 3 or 4 inches in diameter and installed not more than five feet downstream from the first sewer lateral. The park sewer main should be re-vented at intervals of not more than 200 feet.

Cleanouts are installed at intervals of no more than 75 feet and should be the same nominal size as the pipe it serves but no larger than 6 inches.

Sewer laterals should terminate at least 12 inches outside the left wheel and within the rear third of the trailer coach.

In trailers that are properly trapped and vented, the lateral should terminate with a sweep into which is caulked a 3- or 4-inch sanitary tee which terminates 4 to 6 inches above grade. A cleanout should be caulked into the top of the sanitary tee as shown in Figure 5-3. Trailers not properly trapped and vented should have a lateral that terminates with a 3- or 4-inch P-trap into which is caulked a sanitary tee terminating 4 to 6 inches above grade. A cleanout should be caulked in the top of the sanitary tee as shown in Figure 5-4. Since a P-trap is required at the end of the lateral (branch line), the measured horizontal distance from a vented sewer without a re-vent should not exceed 15 feet. The P-trap must not be placed more than 24 inches below grade unless specifically allowed by local authorities.

Water Distribution System

The materials and installation methods for a trailer park water distribution system are the same as for water service piping below grade.

A branch service line connected to the park main supplies potable water to each trailer site. The line terminates on the same side of the trailer site as the trailer sewer lateral.

The minimum size pipe in the park's water distribution system is 3/4 inch. The service connection to each trailer must be a minimum of 1/2 inch.

Trailers are connected to the park's water distribution system with a separate shutoff valve and a springloaded, soft-seat check valve on each branch service line. The valve must be located near the service connection for each trailer.

Size of Sewer (Inches. Based Upon Slope Pitch of ⅛" Per Ft.	Maximum Number of Trailers, Individually Vented System	Maximum Number of Trailers, Loop or Circuit Vented Vented System
3	2	0
4	2	12
5	42	25
6	80	55
8	175	166
10	325	270

Note: This table is usually labeled 46-Y in most code books.

**Installation Method
Table 5-5**

6

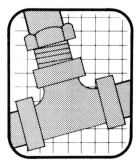

Fire Protection Systems

Fire protection equipment is an important part of many buildings. You have to know the code requirements for proper installation before you can bid these systems.

Standpipe Systems

Many larger buildings must have emergency fire hose connections on the site so firemen will have an adequate supply of water immediately available in the event of a fire. The hose connections are called *standpipes*. Standpipe systems can connect to a public water main as shown in Figure 6-1. Sometimes the public water main can not provide the quantities and pressures required by code. In this case the system must be pressurized. A fire pressure pump is used as shown in Figure 6-2.

Standpipe Requirements

Where standpipes are required, the system must be pressurized ("wet") with a primary water supply constantly or automatically available at each hose outlet.

The estimator should be aware that the standpipes must always be available for fire department use as construction progresses. A fire department connection must be provided on the outside of the building at the street level and at each floor level to the highest constructed floor. Caps should be available to plug standpipe risers in case of an emergency. Each riser must be capped at the end of the day's work. Then the system must be pressurized to the public water main's normal working pressure. This means that additional labor must be figured into the job.

A fire standpipe system is required in buildings over 50 feet high. This means that every building seven stories or more must have standpipes assuming an 8 foot floor-to-floor height.

Buildings designed for theatrical, operatic or similar performances must have a 2½-inch diameter standpipe on each side of the stage. A hose not over 75 feet long must be located at each standpipe hose station.

Standpipe locations must be arranged so that they are protected from mechanical and fire damage. The number of standpipes and hose stations is determined in the following way. All parts of all floors must be accessible to a stream of water and must be within 15 feet of the nozzle end of the hose when a hose not over 100 feet long is connected to the standpipe.

Standpipes must be located within an enclosed stairway. If the stairway is not enclosed, the standpipe must be within 10 feet of the floor landing of an open stairway. Valve or hose connections can not be behind any door.

Sometimes additional standpipes and hose stations are required to provide protection on each floor. Additional stairways may not be provided as they may not be required. In this case, the extra standpipes and hose stations on each floor can be located in hallways or other accessible locations approved by local code authorities.

Estimating Material and Installation

Underground fire lines must have poured concrete

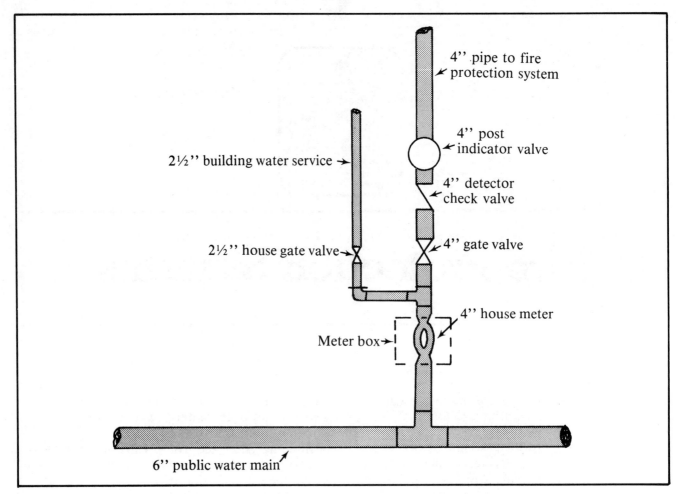

**Typical Connection to Public Water Main Serving Building
Domestic and Fire Protection Needs
Figure 6-1**

thrust blocks resting on undisturbed soil for each change in direction. Figure 6-3 shows various types of thrust blocks. They prevent the pipe and fittings from coming apart under the 200 psi pressure test required for all underground fire lines.

Above-ground fire lines within the exterior wall of a building must be black steel, hot-dipped galvanized steel or copper pipe. The only above-ground fire line is the standpipe. These lines, fittings and all connections must be able to withstand 100 psi pressure at the topmost outlet.

Buildings 50 to 75 feet high must have 4-inch standpipes. Buildings more than 75 feet high must have 6-inch standpipes. Standpipes in buildings 50 feet or higher must extend above the roof a minimum of 30 inches. This extension must have the same diameter pipe as the rest of the standpipe. An Underwriters' Laboratory approved duplex roof manifold with 2½-inch fire department connections must be installed on each standpipe.

Standpipes located in stairway enclosures must also have a 2½-inch fire department outlet. These outlets must have 2½-inch valves adapted for 2½-inch National Standard Thread fire department hose connections. Outlets must be accessible on each floor at the stair enclosure and at the stair enclosure in the basement, if one exists. Hose outlets must not be located farther than 10 feet from the standpipe or hose station, must not be installed lower than 5 feet or higher than 6 feet above the finished floor.

A hose station must be provided within 10 feet of the standpipe. The pipe connecting the hose station to the standpipe is usually 2½-inch pipe. Hose stations must never be located within any stair enclosure. Figure 6-4 shows a typical standpipe layout.

An approved wall-mounted hose reel, cabinet or rack must be provided for each hose station. The station must be located so that it is accessible at all

Fire Protection Systems 61

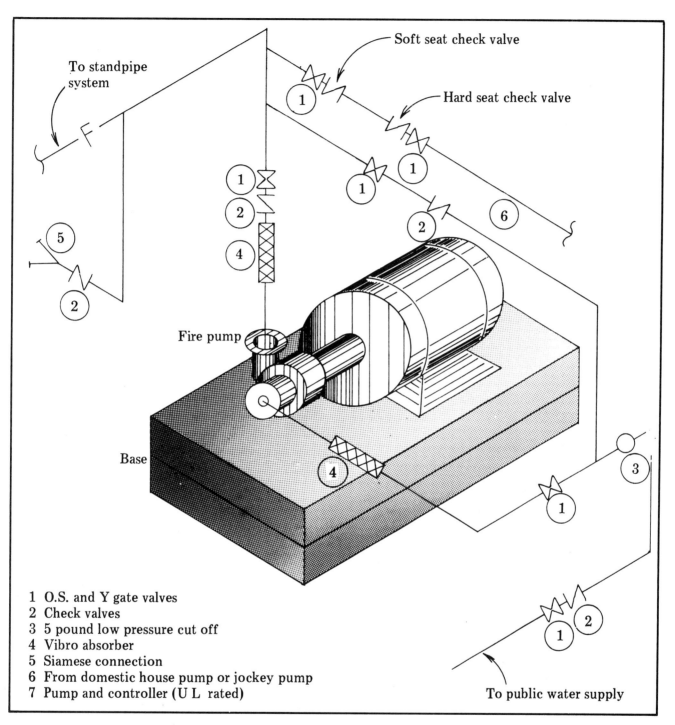

1 O.S. and Y gate valves
2 Check valves
3 5 pound low pressure cut off
4 Vibro absorber
5 Siamese connection
6 From domestic house pump or jockey pump
7 Pump and controller (U L rated)

Fire Pump Detail
Figure 6-2

times. Each hose must be able to withstand 100 psi working pressure and be equipped with an adjustable nozzle that can be used to turn the water on and off. The nozzle should be adjustable from a fog spray to a strong stream of water. Where the pressure may exceed 100 psi, pressure reducers must be installed to control the pressure to the hose.

When more than one standpipe is required to serve a building, each standpipe must be interconnected at its base.

A public water supply main to serve a standpipe system must not be smaller than 4 inches in diameter.

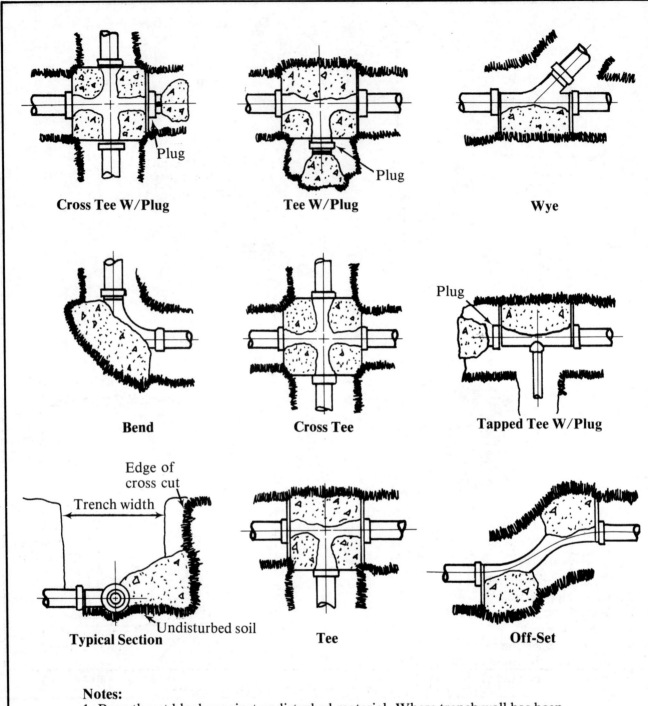

Notes:

1. Pour thrust blocks against undisturbed material. Where trench wall has been disturbed, excavate loose material and extend thrust block to undisturbed material.

2. On bends and tees, extend thrust blocks full length.

3. Place board in front of all plugs before pouring thrust block.

4. In back filling, any muck encountered shall be removed and replaced with acceptable material.

Thrust Blocks
Figure 6-3

Fire Protection Systems

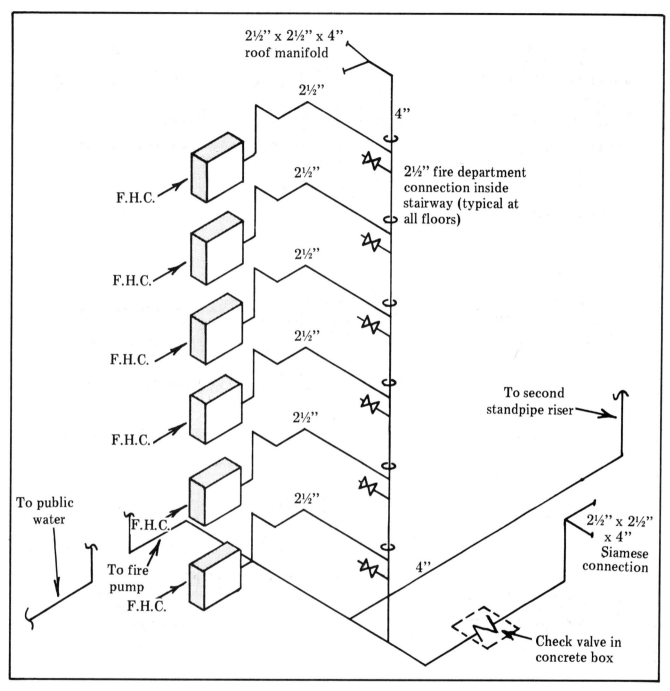

**Standpipe Layout and Location of Fire Hose Cabinets
Figure 6-4**

A Siamese (duplex) fire department connection must be provided for each of the first two required standpipe risers. If more than two standpipes are required, the Siamese connections must be remotely located. Each Siamese connection must have the same diameter as the largest standpipe. For example, a building requiring a 4-inch standpipe must have a 4-inch Siamese connection; a 6-inch standpipe must have a 6-inch Siamese connection.

An approved Underwriters' Listed check valve must be installed between the Siamese connection and the fire standpipe system. Each Siamese connection must have a 2½-inch National Standard Thread fire department hose connection, and must be installed on the street side of the building at least 1 foot but no more than 3 feet above grade.

The Siamese connection or its related piping must not project over public property (such as

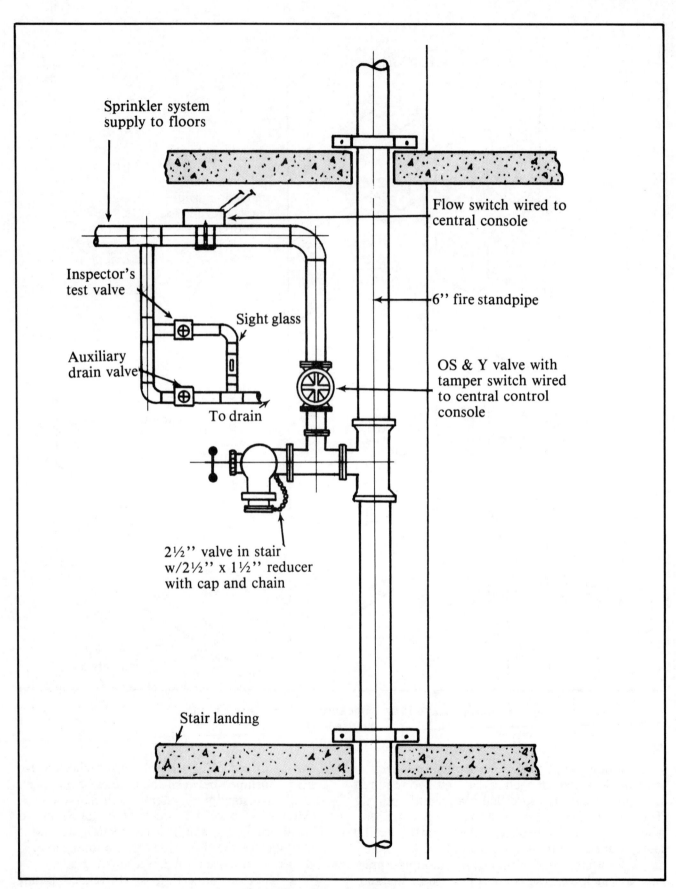

Fire Standpipe Detail
Figure 6-5

Fire Protection Systems

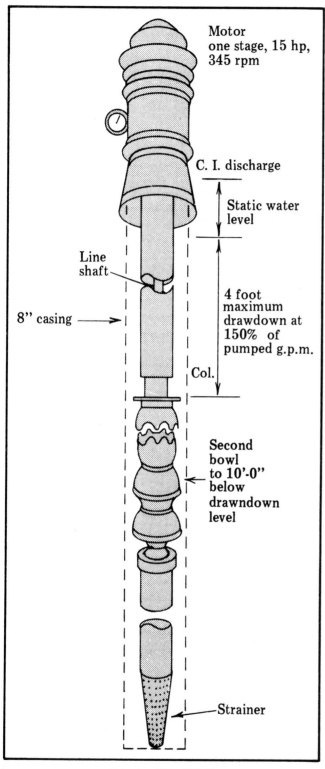

On-site Fire Well
Figure 6-6

sidewalks) more than 2 inches.

Near the fire department Siamese connection a permanent "standpipe" sign with letters 1 inch high should be attached to the exterior of the building.

The water supply for standpipes must be sufficient to maintain 65 psi residual pressure at the topmost outlet, giving a required flow of 500 g.p.m. If more than one standpipe serves a building, the required flow must be 750 g.p.m. In most cases additional pressure and water flow must be supplied by fire pumps. Refer back to Figure 6-2.

Fire pumps required to supply the 500 gallons per minute must be UL Listed. The pump controllers must also be UL Listed and may use limited service motors of 30 hp or less. Fire pumps must operate automatically with compatible controls. Pumps must be supplied with a separate electric service or be connected through a separate automatic transfer switch to a standby generator.

Here are the fire pump planning regulations. Refer back to Figure 6-2 to understand these requirements.

• A 15 psi minimum pressure on a standpipe system at the roof must be maintained by a jockey pump actuated by a pressure switch or by connection to a suitable domestic system through two 170 psi check valves. One of these valves must have a soft seat and one must have a hard seat.

• A full size bypass must be provided with an approved gate and check valve.

• The fire pump must have flexibly coupled drivers.

Special Note: Buildings required to have a sprinkler system need not have hose stations as illustrated in Figure 6-4. But each floor must have a sprinkler supply system connected to a fire standpipe as illustrated in Figure 6-5.

Alternate Water Sources

The plumbing contractor has to install an on-site well system if a fire protection system is required and public water is not available. Information on the following pages will help you estimate this installation.

Fire Flow On-Site Well System

Figures 6-6 and 6-7 illustrate a fire flow on-site well system. The following are code requirements:

• All wells must be cased and sized for a flow of 500 g.p.m.

• The well must be of ample diameter and depth, and must be sufficiently straight to receive the pump.

• All casings must have a wall thickness of at least 3/8 inch.

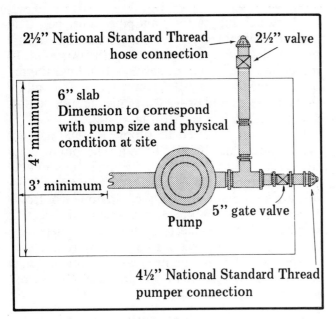

Fire Flow On-site Well—Plan View
Figure 6-7

- The well must be cased to a proper depth and properly sealed to prevent loose or foreign material from entering.
- The well must be properly developed to be free of sand or loose gravel.
- The well must be test-pumped at 150% of the capacity of the pump to be installed for 2 hours after it is free of sand.
- The drawdown in the well must not exceed 4 feet during pumping at 150% of pump capacity as shown in Figure 6-6.
- The flow rate must be 500 g.p.m. at 20 psi where the pump discharges.
- Hook-up must be done according to the requirements of the fire department having jurisdiction. Each fire department connection on the discharge side of the pump must be equipped with a shut-off valve with a diameter not smaller than the size of the discharge opening.
- Fire department connections rated by g.p.m. are required, and must comply with the National

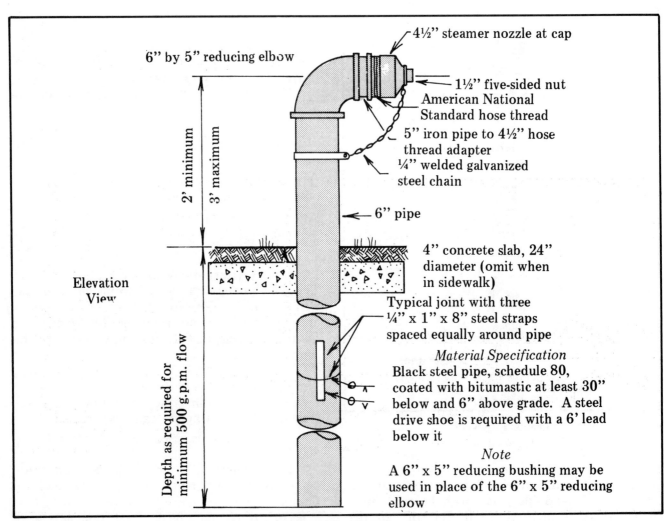

Standard Fire Well Detail
Figure 6-8

Fire Protection Systems

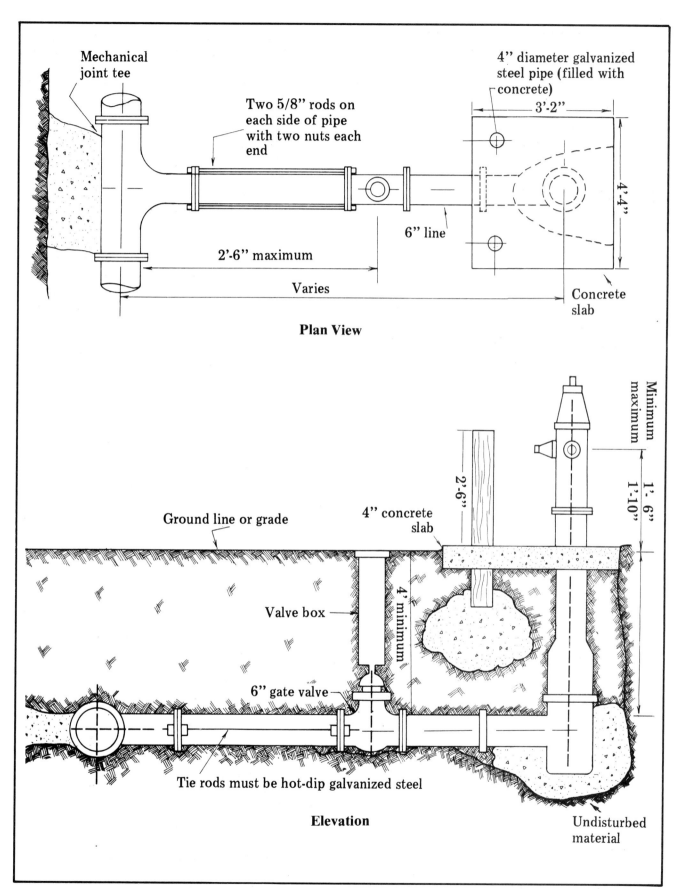

Yard and Street Fire Hydrants
Figure 6-9

Thread Standard (N.T.S.). These connections should have a flow of 500 g.p.m. with one 4½-inch connection and one 2½-inch connection. See Figure 6-7.

- All on-site systems must be tested by the fire department having jurisdiction before final approval of the system.
- On-site systems should be located a minimum of 50 feet from the buildings being protected if physically possible. Fire department access to the on-site system must be by a roadway suitable for fire equipment.
- The fire department connections must be no more than 8 feet from the roadway.
- Direct connections between on-site fire protection systems and the potable water supply are not permitted.

Even where a conventional fire protection system is not required, a standard fire well may be required for certain occupancies if a public water supply is not available. (See Figure 6-8.) The well installation is nearly the same as the on-site well system with the following exception. No pump is required and the hose connection can be a single 4½-inch American National Standard hose connection.

Yard and Street Hydrants

Many plumbing contractors install on-site as well as off-site water distribution systems. Again, as an estimator you should be knowledgeable about such installations.

Yard or street hydrants are usually required in commercial, industrial and residential areas. The code lists other areas that must include yard hydrants because of their size, isolation or type of occupancy. These include mobile home and trailer parks, oil-storage tank yards, lumber yards and exhibition parks. To comply with the code, there must be one yard hydrant and hose station for fire protection for each 20,000 square feet of area in use.

Hydrants must have two 2½-inch connections with National Standard threads similar to those of the local fire department. The fire department must approve the location of all yard hydrants. Figure 6-9 shows a typical hydrant installation.

7

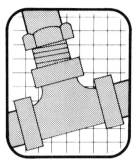

Gas Systems

Plumbing codes seldom, if ever, govern the sizing or installation of gas supply systems. That's because these systems are governed by the local gas code. The gas code is a separate book but is similar to the plumbing code in complexity. Trying to learn about and comply with the gas code can be frustrating and discouraging to even the most experienced plumbing estimator. This chapter will give you a working knowledge of the essential requirements. Information is grouped according to use and is sufficiently simplified for easy application when estimating jobs involving gas systems. However, this information does not replace the gas code. *The gas code is always the final authority.*

The estimator's responsibility for sizing and estimating a building's gas supply system is different from estimating a building's water supply system. The cost of water service piping from the meter (located at the property line) to a building will be included in the contractor's bid price. The cost of gas service piping from its connection with the gas supplier's distribution pipes (located on public property) to the gas meter (generally outside the building) is the responsibility of the gas supplier. The estimator is responsible for the cost of gas piping *only within the building*.

Kinds of Gas

The estimator must have some knowledge of the kinds of gas used in gas systems.

Natural gas piping is the system the estimator will have to work with most often. Natural gas contains chemical impurities which are valuable for uses other than as a fuel. These impurities are removed before being piped to the consumer. The natural gas which millions of consumers use as fuel in homes and industries is known as *dry* or *sweet gas*. Natural gas (methane) is not poisonous but can cause suffocation in a closed space. It is also explosive under certain conditions.

Since natural gas is clean, dry and has no odor, a gas leak might go undetected until there is an explosion. Therefore, a chemical odorant is added before the gas enters the pipeline. This warns of escaping gas before the concentration can reach a dangerous level.

Manufactured gas is produced chiefly from coal. It burns with a blue flame and is generally added to other fuels to increase its heating capacity. Manufactured gas is second to natural gas as a fuel used by consumers in homes and industries. It can be poisonous since it contains carbon monoxide. It is explosive under certain conditions.

Liquified petroleum gas is also known as *LP* or *bottled* gas. It is produced in plants that process natural gas. LPG consists primarily of butane or propane, or a mixture of both. LP gas liquifies under moderate pressure, making it easy to transport and store in special tanks. A LPG tank supplies a building's gas piping system. The liquid becomes a gas again under normal atmospheric pressures and temperatures.

LP gas is heavier than air, colorless, and non-poisonous. Since it is easily containerized and

Length in Feet	Nominal Iron Pipe Size, Inches							
	½	¾	1	1¼	1½	2	2½	
10	176	361	681	1,401	2,101	3,951	6,301	
20	121	251	466	951	1,461	2,751	4,351	
30	98	201	376	771	1,181	2,201	3,521	
40	83	171	321	661	991	1,901	3,001	
50	74	152	286	581	901	1,681	2,651	
60	**67**	**139**	**261**	**531**	**811**	**1,521**	**2,401**	Residential
70	62	126	241	491	751	1,401	2,251	
80	58	119	221	461	691	1,301	2,051	
90	54	111	206	431	651	1,221	1,951	
100	51	104	196	401	621	1,151	1,851	
125	45	94	176	361	551	1,021	1,651	
150	**41**	**85**	**161**	**326**	**501**	**951**	**1,501**	Commercial
175	38	78	146	301	461	851	1,371	
200	36	73	136	281	431	801	1,281	
Column	One	Two	Three	Four	Five	Six	Seven	Eight

More complete sizes will be found in your code.

**Maximum Capacity of Pipe in Cubic Feet of Gas Per Hour
Natural Gas 1,000 B.T.U./Cubic Foot
Table 7-1**

transported, it is convenient to use as fuel for homes and businesses in remote areas.

Sizing Gas Systems

The building gas main and branch lines can be sized if you know the maximum gas demand at each appliance outlet and the length of piping required to reach the most remote outlet. Factors such as pressure loss, specific gravity and gas diversity must be considered, but are already accounted for in the tables in the gas code.

The architect should provide the Btu input rate for each appliance on the blueprints. This shows the maximum gas demand for each type of appliance he plans to use.

Gas appliance manufacturers always attach a metal plate in a visible location on each appliance. This plate gives the maximum input rate in Btu's for that appliance. When used appliances are installed, the Btu rating may not be legible or may be missing. In this case, make sure the appliance inlet pipe is no smaller than the supply pipe. A larger supply pipe will not make the appliance function any better. *The supply pipe should never be smaller than the appliance's inlet pipe, and under no circumstances can it be smaller than 1/2 inch.*

The tables in the gas code prescribe the sizing of gas piping in cubic feet of gas rather than in Btu's. Thus, each Btu input rating has to be converted to cubic feet of gas before sizing the distribution piping.

You can assume that each cubic foot of natural gas releases 1,000 Btu's per hour. The Btu rating of some gas varies from this figure, but using 1,000 Btu per cubic foot is a safe assumption.

Assume you are sizing pipe for a range with a maximum demand of 68,000 Btu's per hour. Divide the value in Btu's by 1,000 to find the demand in cubic feet per hour (c.f.h.). Thus, 68,000 Btu's divided by 1,000 = 68 cubic feet per hour.

Table 7-1 can be used with a sizing method that is quick and easy. *Use it only to help you understand the table in your local gas code.* Low pressure gas systems (natural and manufactured) are the types you will have to work with most often. They are used in millions of home and business installations.

Figure 7-2 shows a gas piping arrangement for a restaurant. This is a typical commercial gas piping system.

Figure 7-3 shows a simple gas piping system similar to what you might find in most single-family residences. Note that each section of piping must be sized to serve the Btu input rating of the appropriate appliance outlet. This illustration and the explanation in this chapter should help you size pipe accurately in any type of low pressure gas system.

Gas Systems

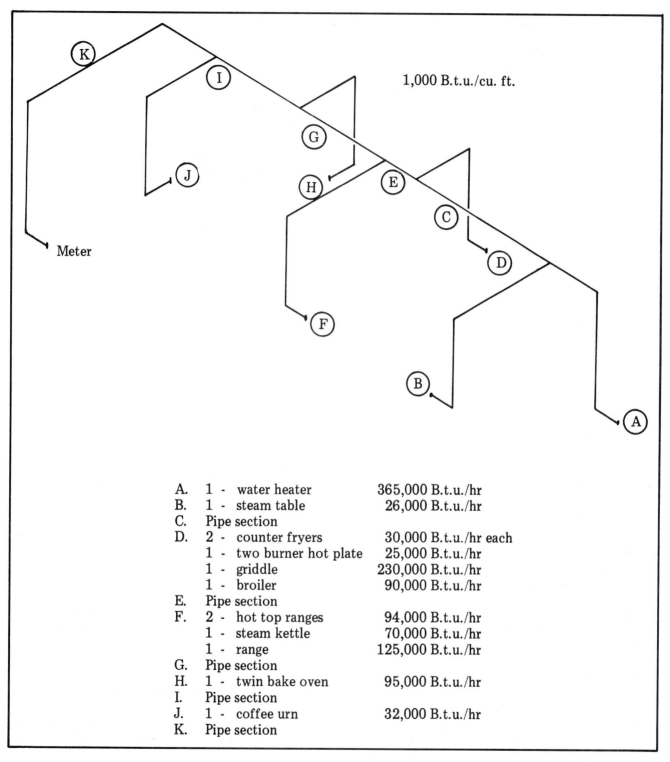

A. 1 - water heater   365,000 B.t.u./hr
B. 1 - steam table   26,000 B.t.u./hr
C. Pipe section
D. 2 - counter fryers   30,000 B.t.u./hr each
   1 - two burner hot plate 25,000 B.t.u./hr
   1 - griddle   230,000 B.t.u./hr
   1 - broiler   90,000 B.t.u./hr
E. Pipe section
F. 2 - hot top ranges   94,000 B.t.u./hr
   1 - steam kettle   70,000 B.t.u./hr
   1 - range   125,000 B.t.u./hr
G. Pipe section
H. 1 - twin bake oven   95,000 B.t.u./hr
I. Pipe section
J. 1 - coffee urn   32,000 B.t.u./hr
K. Pipe section

Natural Gas - Commercial Kitchen Installation
Figure 7-2

Commercial example. To size the gas piping system in Figure 7-2, use Table 7-1 and follow the procedure outlined below. Assume that the total developed length of gas piping in Figure 7-2 is 147 feet from the meter to outlet A. Find that distance in Table 7-1. You have to use the next longer distance if the exact length is not given. *The developed length is the only distance used to determine the size of any section of the gas piping.*

In Table 7-1 (first column), the length of piping

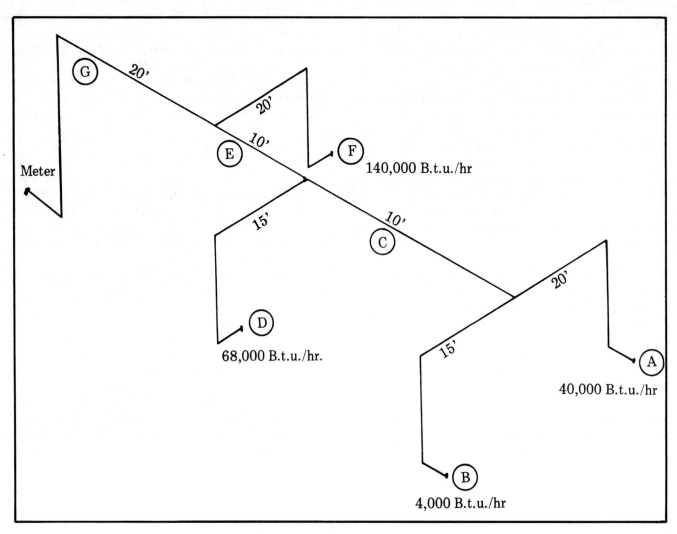

Natural Gas - Residential Installation
Figure 7-3

(147 feet) exceeds 125 feet, but not 150 feet. The figure 150 must be used to determine the size of each section of pipe. It will help if you underline all figures in each vertical demand column opposite 150. (This row is in bold type in Table 7-1.) The pipe size listed at the top of each demand column is the size that will carry the volume listed in that column.

You are now ready to size each section of pipe shown in Figure 7-2. The maximum gas demand of outlet A is 365,000 Btu's per hour, or 365 cubic feet per hour, assuming 1,000 Btu's per cubic foot. Looking across the 150 foot row, the first figure exceeding 365 is found in column 6 (501 c.f.h.). The correct pipe size for section A is 1½ inches.

Use the same procedure to size each section. The maximum gas demand for section B is 26 c.f.h., and the correct pipe size is 1/2 inch (column 2). The maximum gas demand for section C, which includes the combined demands of sections A and B, is 391 c.f.h. The pipe size would be 1½ inches, as the total exceeds 326 in column 5 but not 501 in column 6. The maximum gas demand for section D includes the combined demands of 5 appliances: 30,000 plus 30,000, plus 25,000, plus 230,000, plus 90,000. This converts to 405 c.f.h. Again, using column 6, the pipe size would be 1½ inches. The maximum gas demand for section E, which includes the combined c.f.h. of sections A, B, C and D, is 796 c.f.h. This exceeds 501 in column 6 but not 951 in column 7. This section of pipe must be 2 inches. The maximum gas demand for section F is 289 c.f.h. Note carefully that the 2 top burners have a *combined* Btu rating of 94,000. The correct pipe size for section F is 1¼ inches (column 5). The maximum gas demand for section G, which includes the combined demand of sections A, B, C, D, E, and F, is 1,085 c.f.h. Use the figure 1,501 in column 8. The correct pipe size for section G is 2½ inches.

The maximum gas demand for section H is 95 c.f.h. and the correct pipe size is 1 inch (column 4). The maximum gas demand for section I, which includes the combined demands of sections A, B, C, D, E, F, G, and H, is 1,180 c.f.h. Column 8 is used since the combined total does not exceed 1,501. The pipe size for section I would be 2½ inches. The maximum gas demand for section J is 32 c.f.h., and the correct pipe size is 1/2 inch (column 2). The maximum demand for section K is the cumulative total for the entire restaurant. This is 1,212 c.f.h. The correct pipe sizes would be 2½ inches (column 8). This completes the gas piping installation.

Residential example. Size the gas piping for a single family residence the same way. The total developed length of the gas piping for a residence will be shorter and the Btu ratings of the appliances will be less.

The developed length of gas piping in Figure 7-3 (measured from the meter to the most remote outlet—A in this case) is 60 feet. In Table 7-1 (column 1), locate the figure 60. The pipe size at the top of each demand column is the correct pipe size to use for the volume given.

The maximum gas demand of outlet A is 40,000 Btu's per hour or 40 c.f.h. The correct pipe size is 1/2 inch (column 2). The maximum demand of outlet B is 4,000 Btu's or 4 c.f.h., and the correct pipe size is 1/2 inch (column 2). The combined maximum demand for A and B is 44 c.f.h., and the correct pipe size for Section C is 1/2 inch (column 2). Maximum demand of outlet D is 68,000 Btu's per hour, or 68 c.f.h., and the correct pipe size is 3/4 inch (column 3). The combined maximum demand for pipe section E is 112 c.f.h. The correct pipe size is 3/4 inch (column 3). The maximum demand of outlet F is 140,000 Btu's per hour, or 140 c.f.h., and the correct pipe size is 1 inch (column 4). The combined maximum gas demand for pipe section G is 252 c.f.h. This is the maximum gas demand for the entire residence. The correct pipe size is 1 inch (column 4).

It's a good idea to make a piping diagram before estimating a gas system. The Btu rating of each appliance should be noted for each outlet, as illustrated in Figure 7-2 and 7-3. Each section of pipe should then be sized as shown above.

Enter the pipe sizes, fittings, valves and other miscellaneous materials from the blueprints on your material take-off sheet for pricing. The linear footage of piping and number and type of fittings will determine the man-hours necessary for the installation. Man-hour production tables are included in Chapter Sixteen.

8

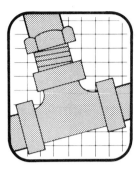

Sanitary Sitework & Water Distribution Systems

This chapter provides an in-depth view, with illustrations, of special systems every estimator encounters. The plumbing systems and components in this chapter are based on the quantities of materials in the drawings in Part III, Chapter Twelve.

Sanitary Sitework Systems

For most small jobs the sanitary sitework consists of a simple building sewer. The building sewer begins at its connection to the public sewer a few inches within the property line. It terminates at its connection to the building drain approximately five feet from the building. See Figure 8-1.

Large sanitary installations include the collection system for one or more buildings. This includes all piping, manholes, sewage ejectors and lift stations necessary to convey sewage from one or more building drain lines to an approved method of disposal. In most cases the sanitary collection system is the general contractor's responsibility. The plumbing contractor should be qualified and many times is expected to do this type of work. Figure 8-2 shows such a collection system with its components.

Manholes

Most model codes require that manholes be installed as follows (see Figure 8-2):

1) Not greater than 300 feet apart and at the end of each private sewer.
2) At every change of grade, size or alignment.
3) At the connection with a building sewer only where such building sewer is larger than the private sewer.
4) At the connection with a public sewer.

Manholes are provided so the sewer can be entered for cleaning, inspection and repair. They may be constructed of brick at the job site; precast manholes may be delivered and installed as a single unit. Figures 8-3 through 8-10 give detailed requirements for the two standard types of manhole installations.

Connecting Private and Public Sewers

The public sewage collection system is located in either a street, alley or a dedicated easement adjacent to each parcel of private property. Public sewers are common pipes installed, maintained and controlled by local authorities.

During the installation of the sewage collection system, a 6-inch sewer lateral is usually extended from the main to several inches past the property line on each lot. If the lateral extension is not done at the time of the original installation, the plumbing contractor has to make this connection.

The local municipal engineering department can tell you the depth, size and location of their sewer pipe. Visit the job site to see whether sidewalk, curb or pavement might have to be removed to make this connection. The cost for replacing these items should be included in your bid price. A permit must be obtained from the local authority and an inspection will be required for the connection.

Figures 8-11 and 8-12 show standard sanitary sewer connection details accepted by most

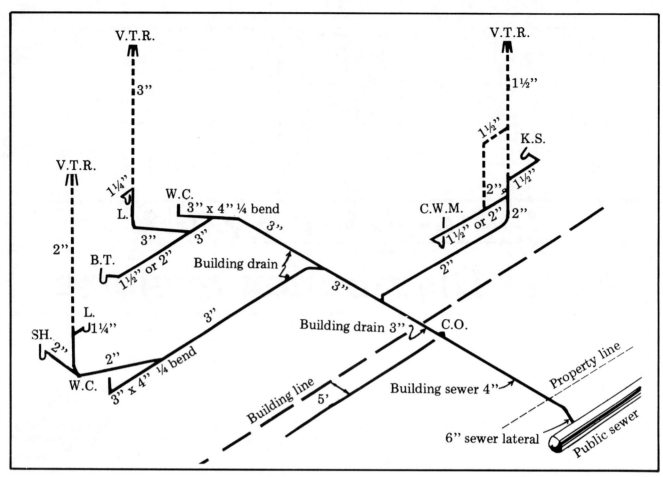

Simple Building Sewer Pipes
Figure 8-1

municipal engineering departments. Figure 8-11 is a shallow connection and Figure 8-12 is a deep connection.

Sewage Lift Stations and Force Mains
Sewage lift stations are required when a building drainage system is lower than the public sewage collection system. See Figure 8-2. The building sewage flows into the lift station wet well. Pumps and wet well must be able to handle the peak daily flow as specified in the design criteria. (See Figure 8-14.) The sewage is lifted by alternating pumps and forced under pressure through the force main into the public sewage collection system. Complete details of a typical lift station and force main installation are shown in Figures 8-13 and Figures 8-15 through 8-19.

Sewage Ejectors
When plumbing fixtures or appliances are installed below the crown of the street, they discharge by gravity into a receiving tank. Then the waste must be lifted and discharged into the building sewer or drain by ejectors. The receiving tank must retain a 30-minute peak flow. Pump discharge pipes must be provided with a check located as close as possible to the pump and on the pump side of a gate valve.

A single ejector pump may be used for one- or two-family buildings. Duplex ejector pumps must be used in all other buildings. Receiving tanks collecting sewage must have a securely fastened gas and air tight metal cover. An air and gas-tight manhole is required for repair access. The receiving tank should have a minimum 3-inch vent. Figure 8-20 shows a single ejector pump.

Sump Pump
In places where clear water liquid waste drainage is required and the floor level is below the building drainage system, the waste must discharge into a receiving sump. A receiving sump which collects clear water liquid waste need not be vented or covered. A typical clear water liquid waste sump and pump detail is shown in Figure 8-21.

Sanitary Sitework & Water Distribution Systems

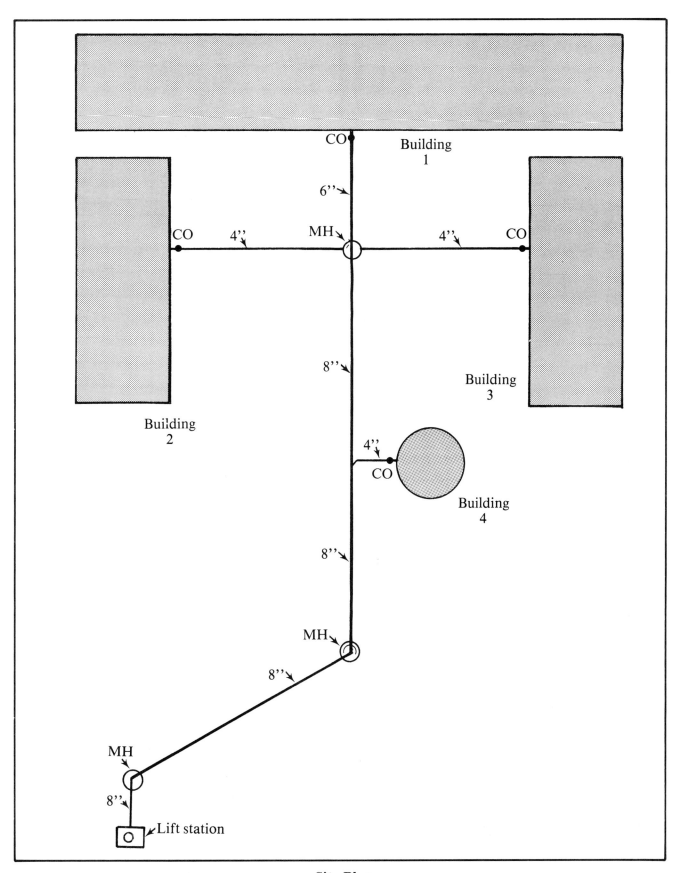

**Site Plan
Sewage Collection System for 42 Unit Apartment Complex
Figure 8-2**

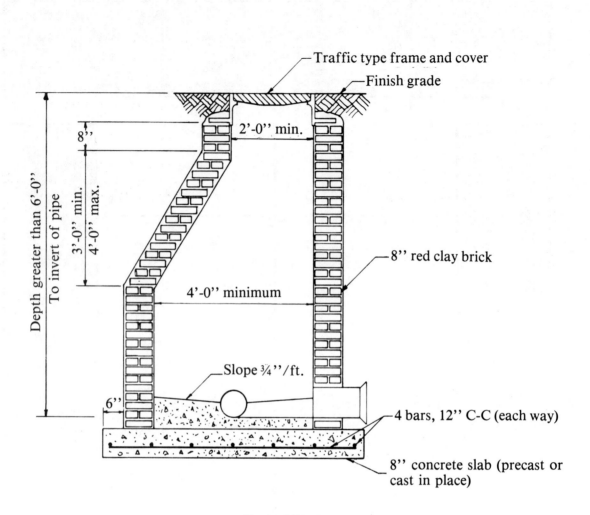

Typical Section

Notes:
1. Brick masonry construction to be stuccoed with ¾" mortar inside and outside.

2. The first three sections of pipe at both influent and effluent of each manhole to be a maximum of 2'-0" in length.

3. A flow channel shall be constructed inside manhole to direct influent into flow stream.

**Standard Sanitary Sewer Detail
Standard Manhole (Brick)
Figure 8-3**

Sanitary Sitework & Water Distribution Systems

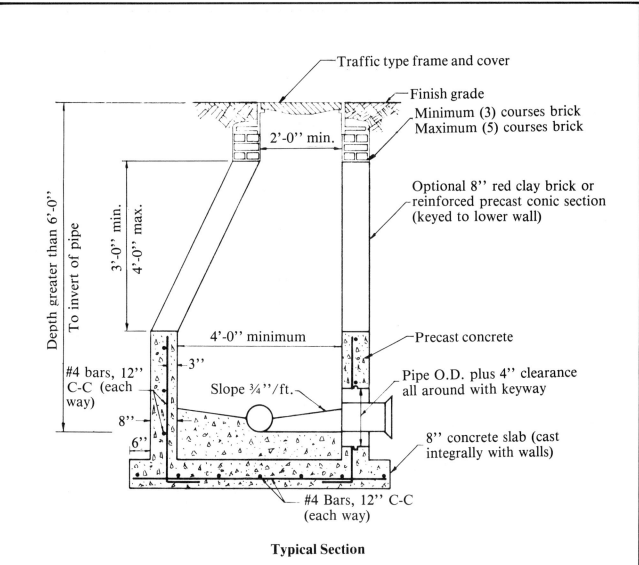

Typical Section

Notes:
1. Brick masonry construction to be stuccoed with ¾" mortar inside and outside.
2. The first three sections of pipe at both influent and effluent of each manhole to be a maximum of 2'-0" in length.
3. Lift holes through precast structure are not permitted.
4. See technical specifications for placement of construction joints.
5. All openings shall be sealed with a waterproof, expanding grout.
6. A flow channel shall be constructed inside manhole to direct influent into flow stream.

Standard Sanitary Sewer Detail
Standard Manhole (Precast)
Figure 8-4

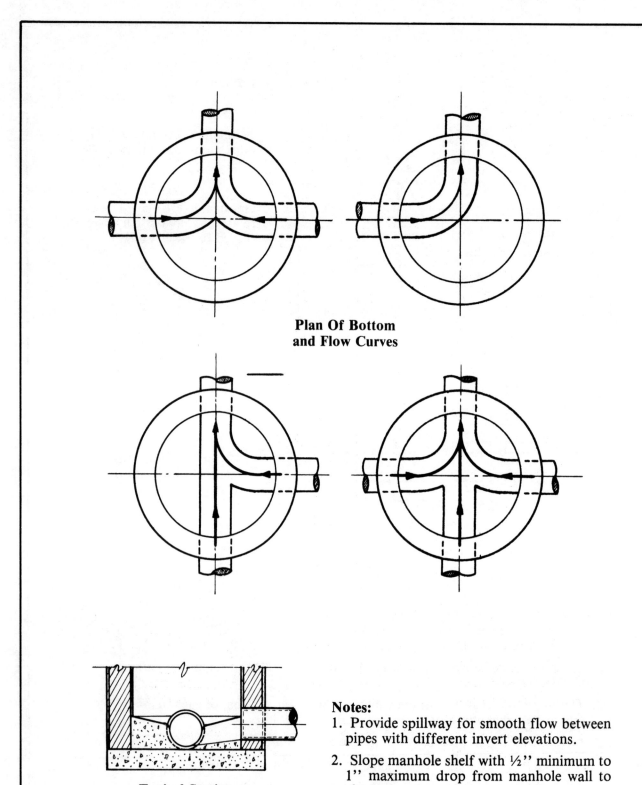

Standard Sanitary Sewer Detail
Flow Patterns for Bottom of Manhole
Figure 8-5

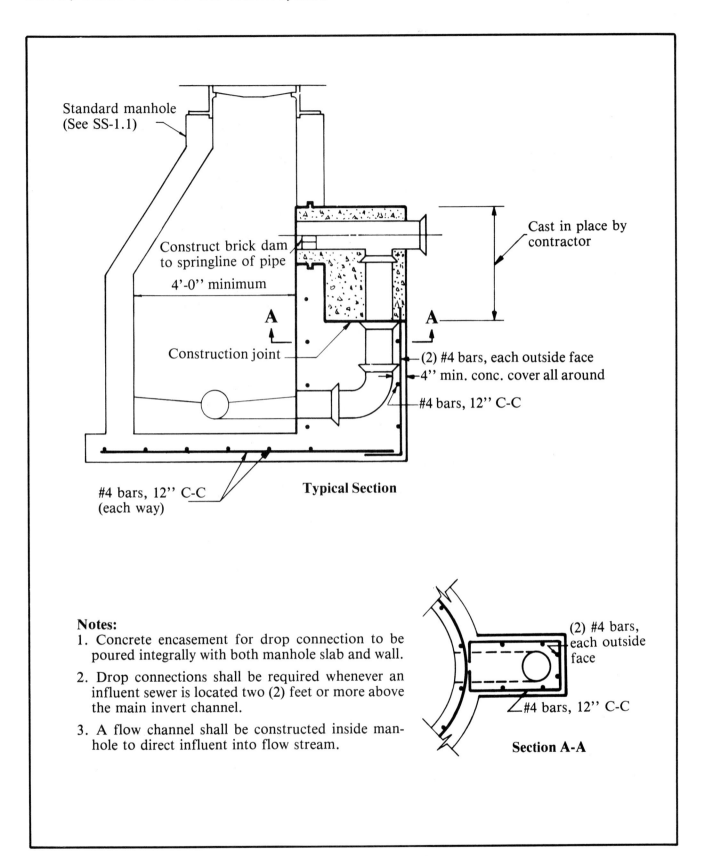

**Standard Sanitary Sewer Detail
Drop Connection (New Precast Manhole)
Figure 8-6**

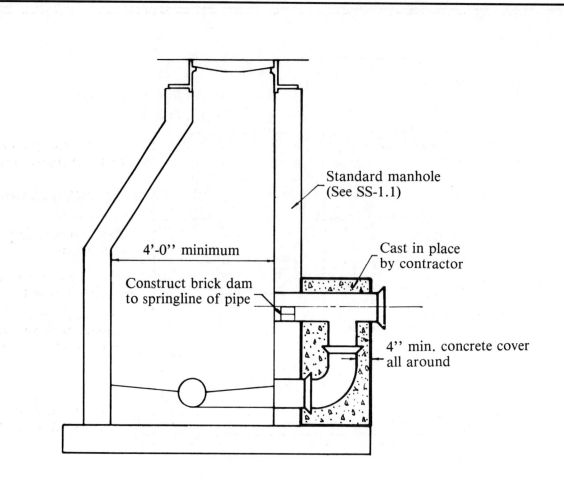

Typical Section

Notes:
1. Drop connections shall be required whenever an influent sewer is located two (2) feet or more above the main invert channel.
2. A flow channel shall be constructed inside manhole to direct influent into flow stream.
3. Construction of brick manholes shall provide an oversized slab to extend under the drop connection.

**Standard Sanitary Sewer Detail
Drop Connection (New Brick Manhole)
Figure 8-7**

Sanitary Sitework & Water Distribution Systems

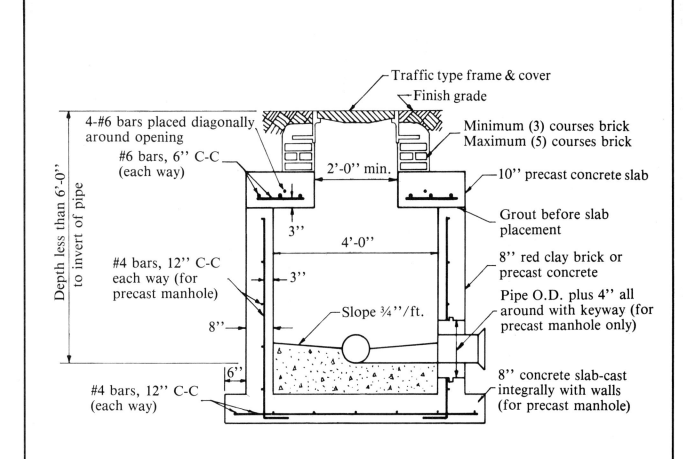

Typical Section

Note:
Brick masonry to be stuccoed with ¾" mortar inside and outside.

2. The first three sections of pipe at both influent and effluent of each manhole to be a maximum of 2'-0" in length.

3. Lift holes through precast sections not permitted.

4. All openings shall be sealed with a waterproof, expanding grout.

5. A flow channel shall be constructed inside manhole to direct influent into flow stream.

**Standard Sanitary Sewer Detail
Shallow Manhole (Brick or Precast)
Figure 8-8**

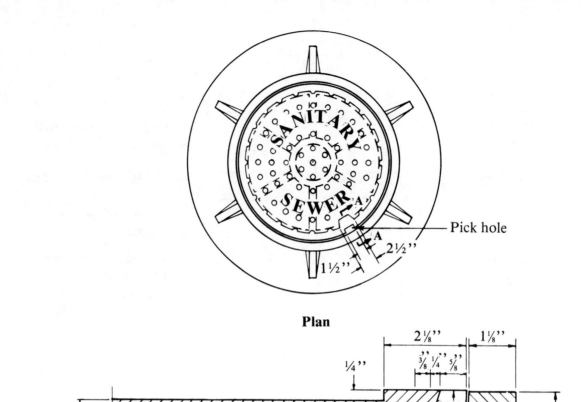

Plan

Section A-A

Notes:
1. Letters on cover to be arranged with a radius of 6½" to center of letters.
2. Each letter to be 2" long, ⅜" deep, ¼" to 5/16" thick, and flush with top of beads.
3. Beads to be ⅜" high with a radius of ½" at bottom and ⅜" at top.
4. All bearing surfaces to be machined.
5. Minimum weights:
   Cover: 168 lbs.
   Frame: 365 lbs.

**Standard Sanitary Sewer Detail
Sanitary Sewer Manhole (Frame and Cover)
Figure 8-9**

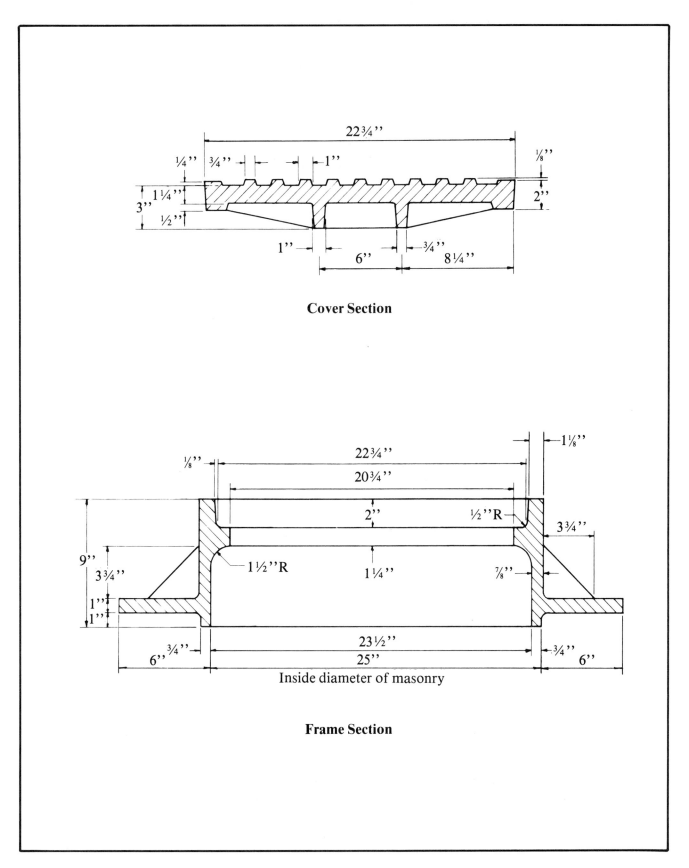

**Standard Sanitary Sewer Detail
Sanitary Sewer Manhole (Frame & Cover)
Figure 8-10**

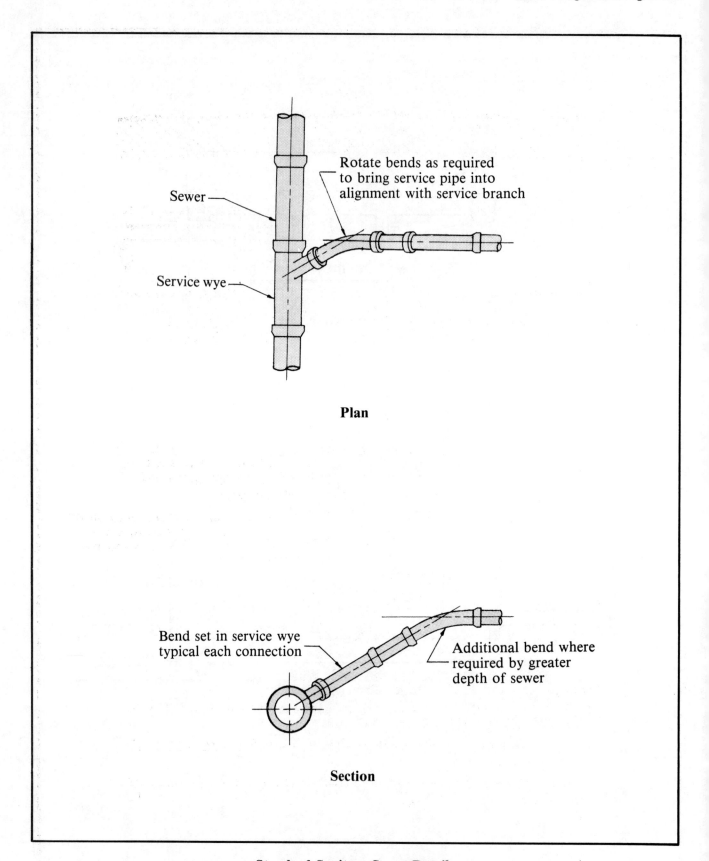

**Standard Sanitary Sewer Detail
Shallow Connection (Where Invert of Sewer is Less Than 6'-0" Deep)
Figure 8-11**

Sanitary Sitework & Water Distribution Systems

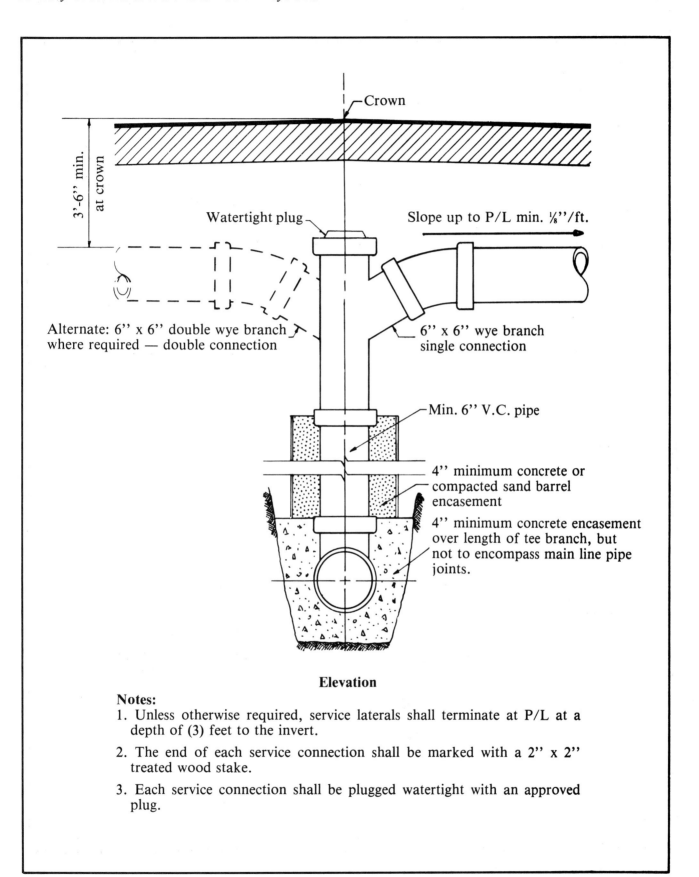

**Standard Sanitary Sewer Detail
Typical Riser Connection (Where Top of Sewer is 7'-0" or Deeper)
Figure 8-12**

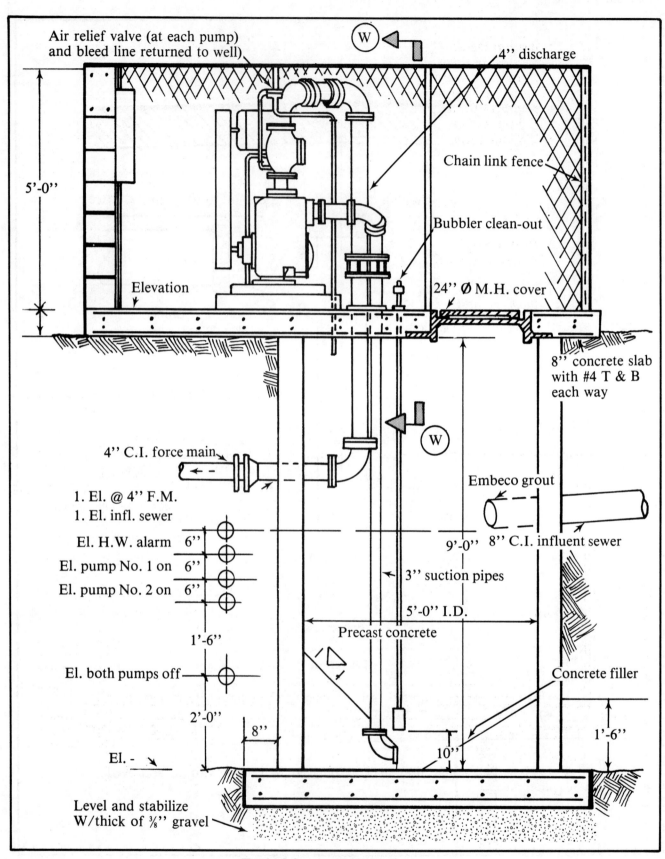

Typical Sewage Lift Station
Section (V) - (V) (Scale: ½" = 1'-0")
Figure 8-13

1. **Design Data:**
   Number of units -- 42
   Number of bedrooms -- 126
   Gallons per day per bedrooms --- 150
   Number of washing machines --- 42
   Gallons per day per washing machine ---------------------------------- 210
2. **Computations:**

   $$126 \times 150 = 18,900 \text{ G.P.D.}$$
   $$42 \times 210 = 8,820$$
   $$\text{Total} = 27,720 \text{ G.P.D.}$$

   $$\text{Average flow} = \frac{27,720}{24 \times 60} = 19.25 \text{ G.P.M.}$$

3. **Peak Flow:**
   $2.5 \times 10.5 = 26.25$ G.P.M. plus 10% infl. 2.63 = 29.88 G.P.M.

   To provide a minimum velocity of 2.5 ft./sec. in the 4'' force main, use 100 G.P.M. pumps.

Design Criteria
Figure 8-14

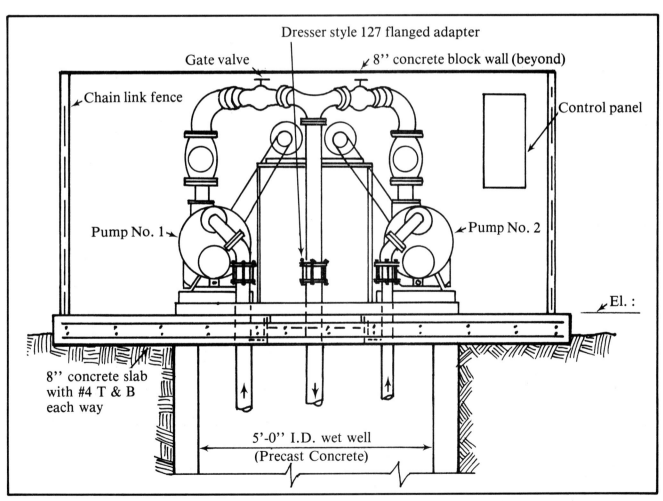

Typical Sewage Lift Station
Part Section (W) - (W) (Scale: ½'' = 1'-0'')
Figure 8-15

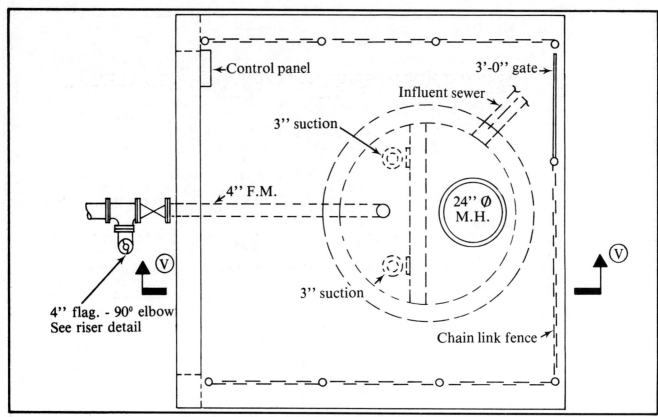

Part Plan (Pumps Not Shown) (Scale ½" = 1'-0")
Sewage Lift Station
Figure 8-16

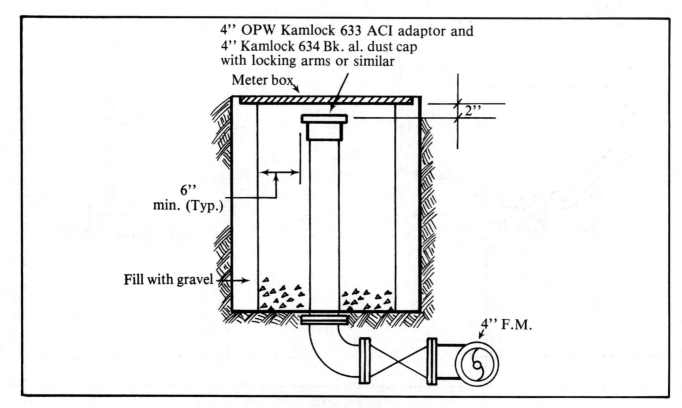

Emergency Connection Riser Detail (A)
Figure 8-17

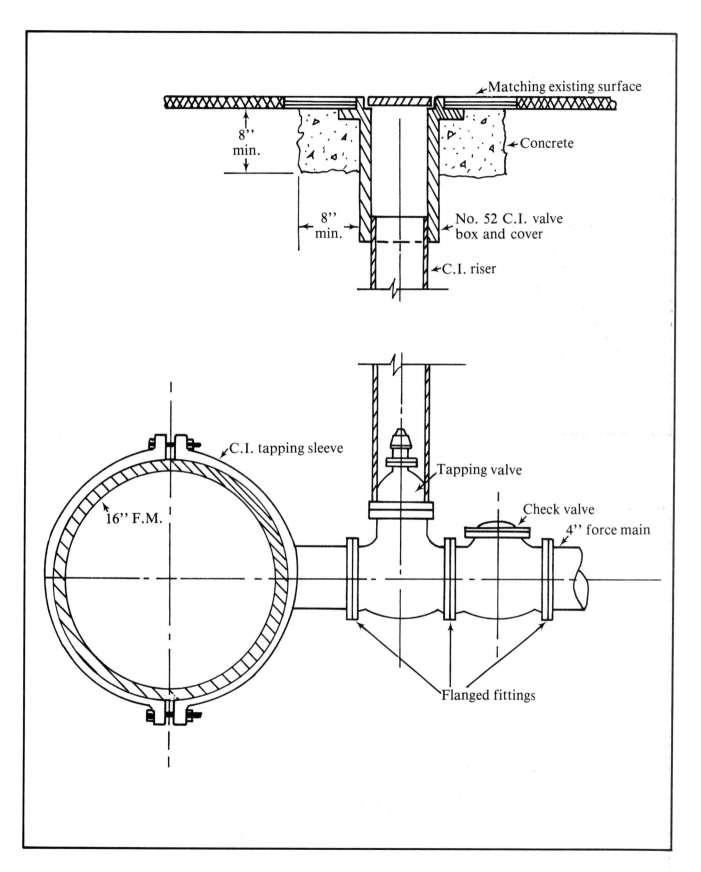

Detail of Force Main Connection (B)
Figure 8-18

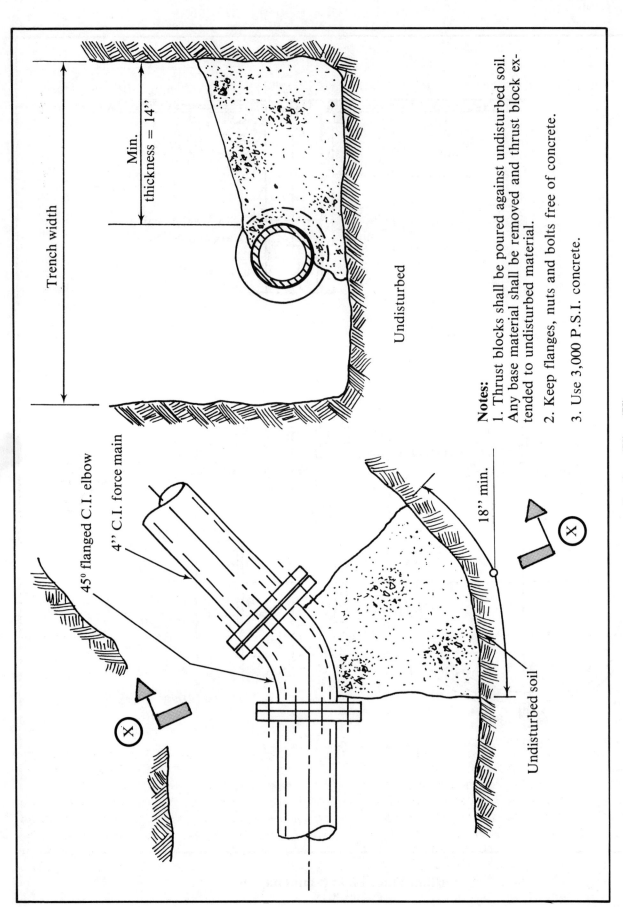

Section (X) - (X)
Plan Detail (C): Thrust Block
Figure 8-19

Sanitary Sitework & Water Distribution Systems

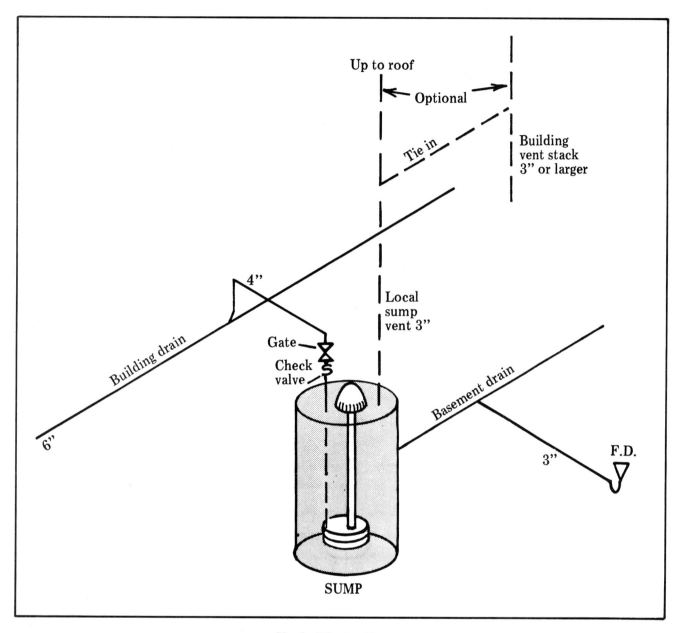

Single Ejector Pump
May be Used in One or Two Family Buildings
Figure 8-20

Transformer Vault Drainage

Transformer vault rooms within a building must have floor drainage. A 3-inch floor drain (less trap) is installed at floor level. The drain pipe connects to an oil spill holding tank located outside the building. The drain line does not have to be vented. The oil spill holding tank is sized by the power company, which also determines the amount of oil contained in the transformers being used. Figure 8-22 is a detail of a typical transformer oil spill holding tank and piping.

Trap Reseal Detail

Floor drains connected directly to the sanitary drainage system have a water seal trap. The code requires that all traps serving floor drains have a constant source of water to maintain an adequate water seal. This is to prevent odors from entering the building. An automatic trap resealer must be installed, as shown in Figure 8-23, where a constant source of water is not available, as in public bathrooms.

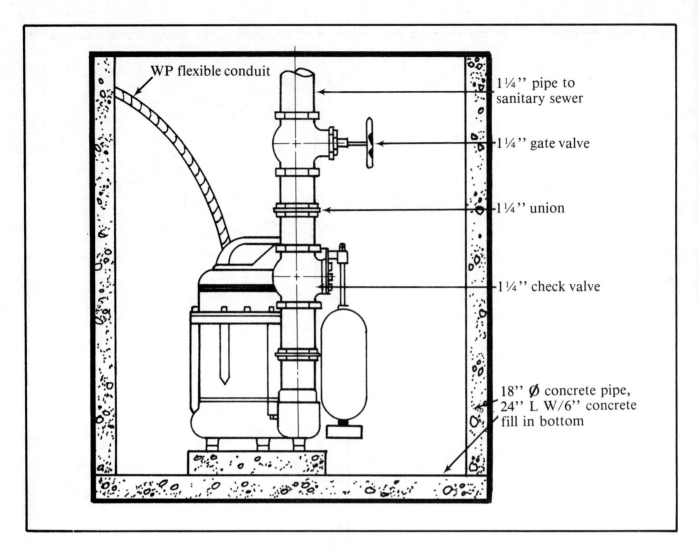

Sump Pump Detail (N.T.S.)
Figure 8-21

Settling Tank, Drainage Well Detail

Storm drainage from large complexes must be properly collected and disposed of. If storm sewers are not available, a drainage well is used to legally dispose of rain water collected from the roof and parking decks. The storm drainage system must discharge into a settling tank to filter any objectionable substances (sand, leaves, grease) before the water enters the drainage well. This should eliminate the possibility of drainage well problems. (See Figure 8-24.)

Trash Chute Piping Detail

Adequate drainage and fire protection must be provided in high-rise buildings with trash rooms on the first floor fed by trash chutes from upper floors.

A floor drain with a backwater valve is required on the first floor to prevent sewage from overflowing onto the trash room floor in case of a stoppage. Most codes require that floor drains be installed on each floor in each vestibule if the vestibule size exceeds 15 square feet. Also, a floor drain is required, regardless of vestibule size, if a fire sprinkler head is installed. This is shown in Figure 8-25.

The trash room must have an adequate number of fire sprinkler heads to protect it from fire. The fire sprinkler heads must be located on every other floor within the trash chute. A flushing ring must be provided above the highest floor level for flushing and cleaning the chute manually, when required. (See Figure 8-25.)

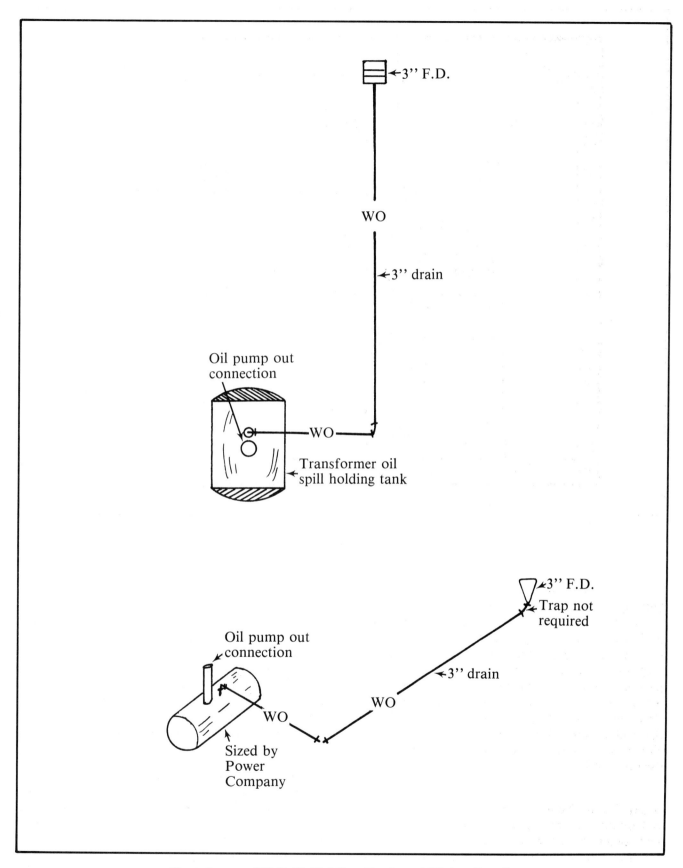

Transformer Oil Spill Holding Tank and Piping Detail
Figure 8-22

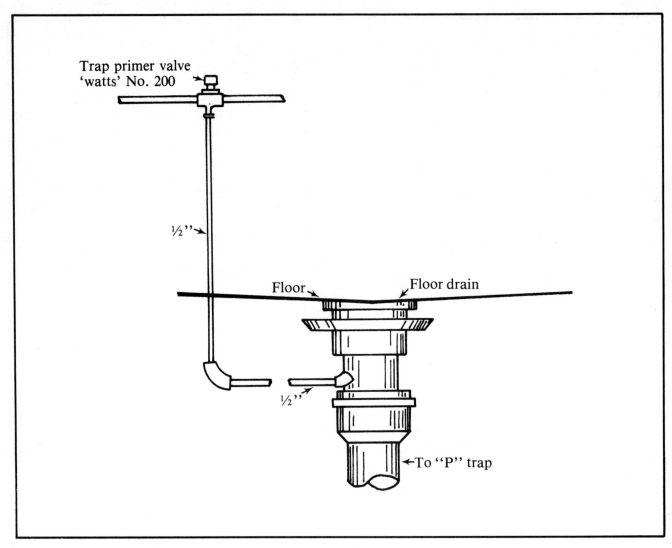

Trap Reseal Detail
Figure 8-23

Accessory Building and Additions
Most codes make an exception for accessory buildings and additions to residential buildings that are located on the same lot with an existing building that has a single building sewer. The sewer from the accessory building (Figure 8-26) and the addition (Figure 8-27) may be sized as a building drain (3 inches).

Horizontal Wet Vents
This special method of venting is generally used in dwellings. It provides adequate protection for trap seals for permissible plumbing fixtures. Being a single piping system, it costs less and can vent several adjoining fixtures when located on the same floor level. (See Figure 8-28.) Check your local code for restrictions, pipe sizes and maximum capacities for wet vents.

Combination Waste and Vent
This special method of receiving waste and venting certain plumbing fixtures is generally used in high-rise buildings. It can be used to advantage in any multi-story building where plumbing fixtures are located directly above each other on different floor levels. This is a single vertical pipe riser installation and is both economical and practical. Figures 8-29 and 8-30 show two types of installation.

The code limits the size of the stack, number of fixture units it may receive, type of plumbing fixtures and the total length of the combination waste and vent. Check your local code requirements when bidding a combination waste and vent system.

Figures 8-31 through 8-34 will help you estimate and design plumbing systems for multi-story buildings.

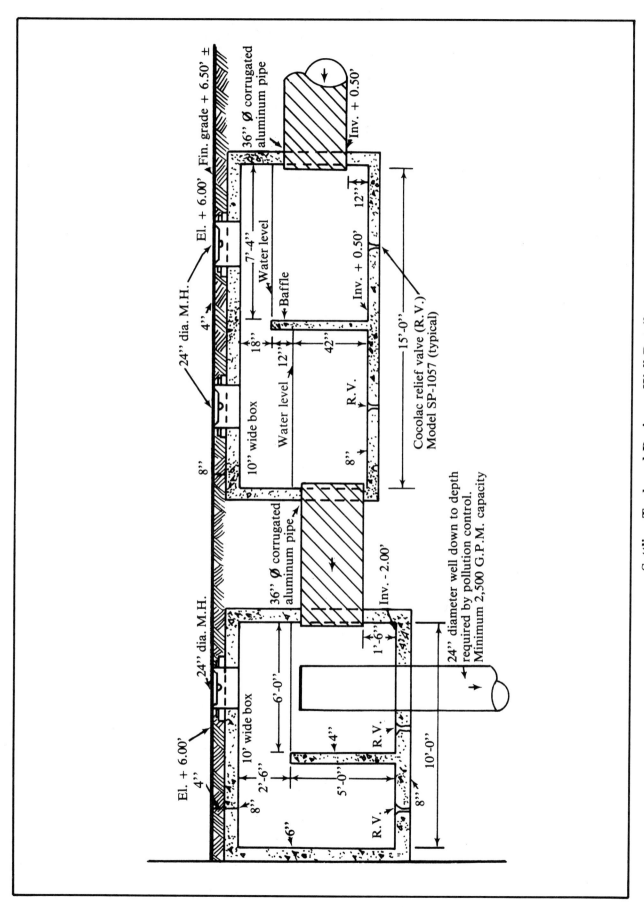

Settling Tank and Drainage Well Detail
Figure 8-24

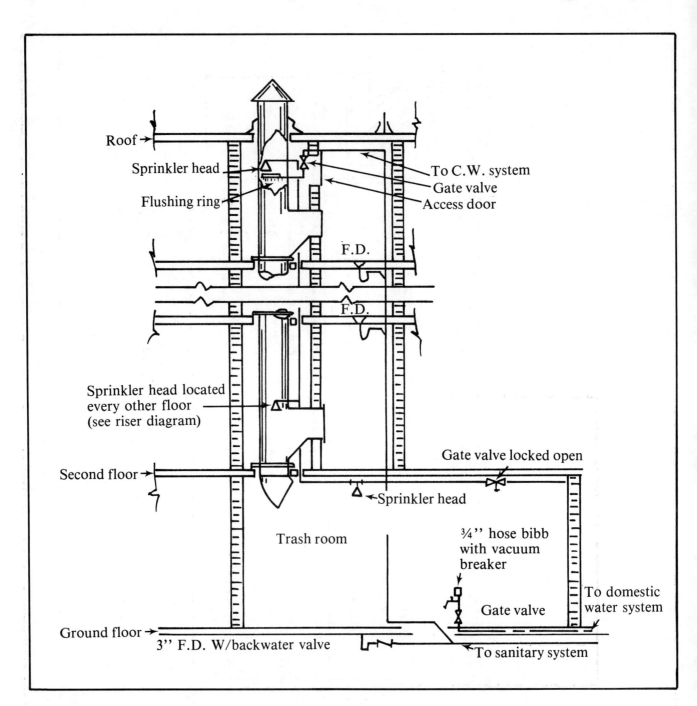

Trash Chute Piping Detail
Figure 8-25

Sanitary Sitework & Water Distribution Systems

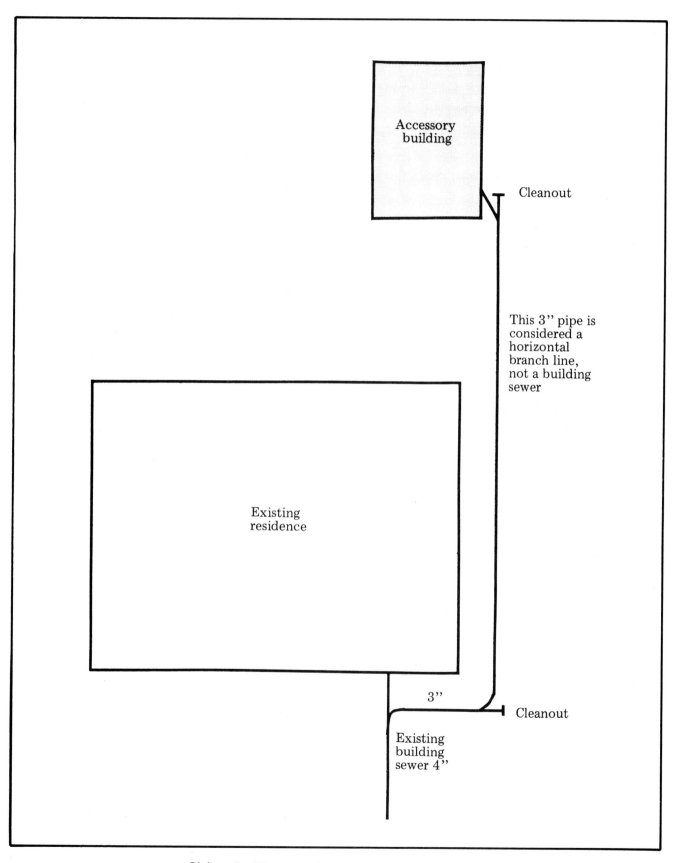

Sizing the Pipes — Accessory Building Hook-up
Figure 8-26

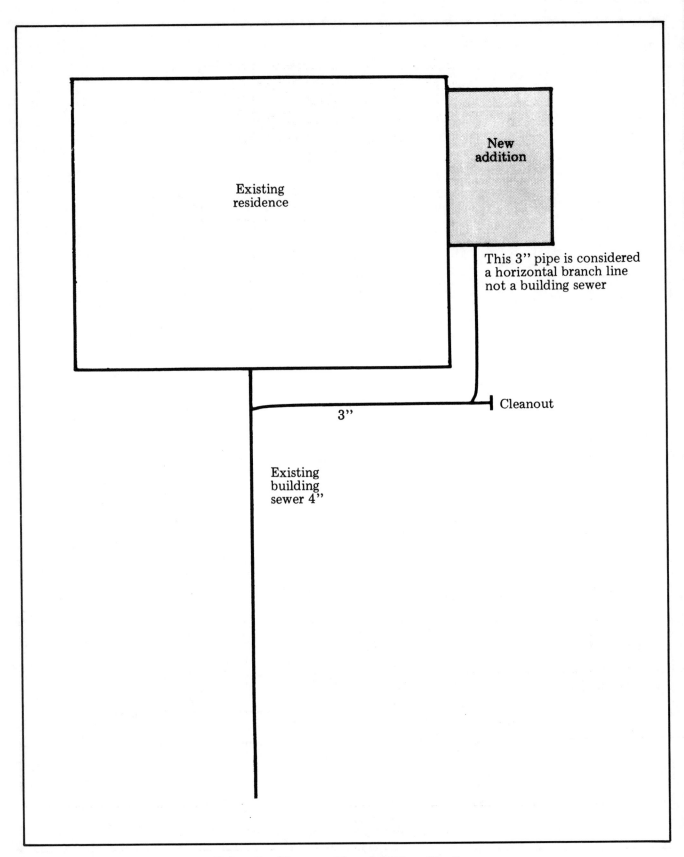

Sizing the Pipes — New Addition Hook-up
Figure 8-27

Sanitary Sitework & Water Distribution Systems

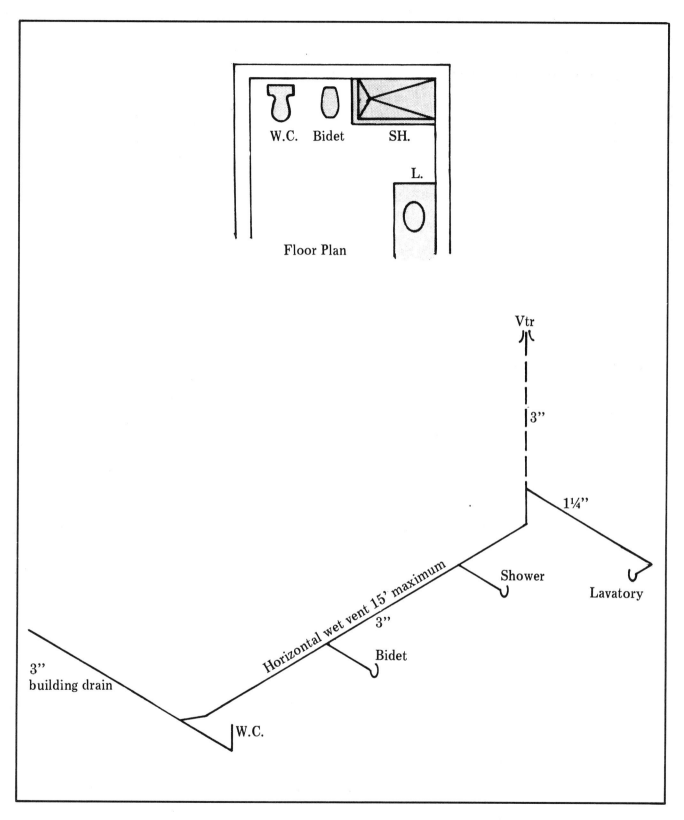

Horizontal Isometric of Wet Vent
Figure 8-28

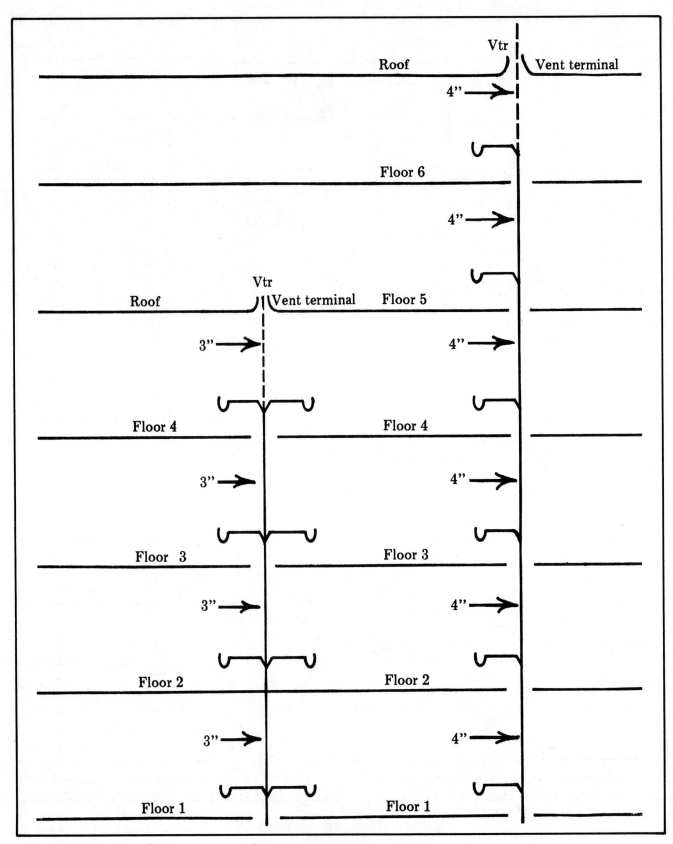

Combination Waste and Vent

Figure 8-29   **Figure 8-30**

Sanitary Sitework & Water Distribution Systems   103

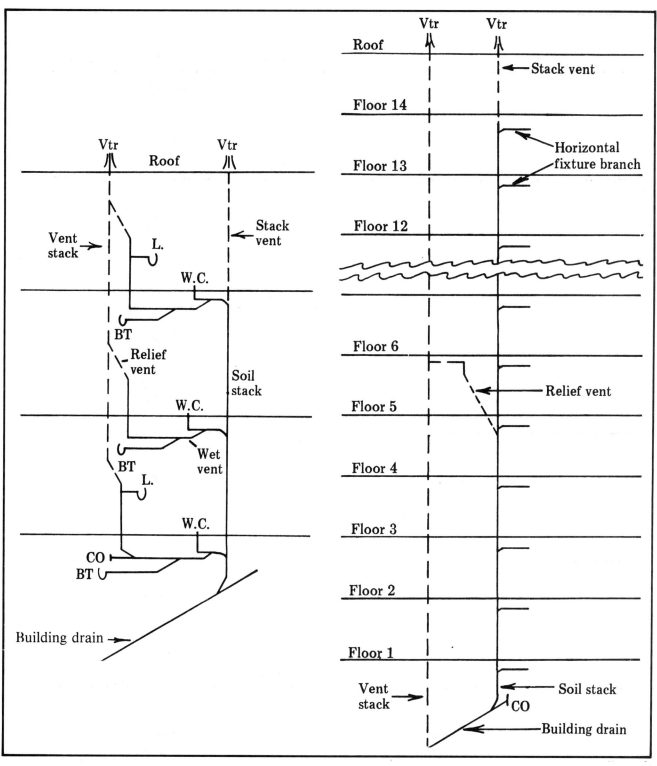

Alternate Method of Wet Venting Single Bathrooms in Multi-Story Building
Figure 8-31

The Diameter of Relief Vent Must be Equal to The Diameter of the Vent Stack it Connects to.

To Balance the Pressures Generated Within a Plumbing System in Multi-story Buildings, You Must Provide a Relief Vent at Various Intervals
Figure 8-32

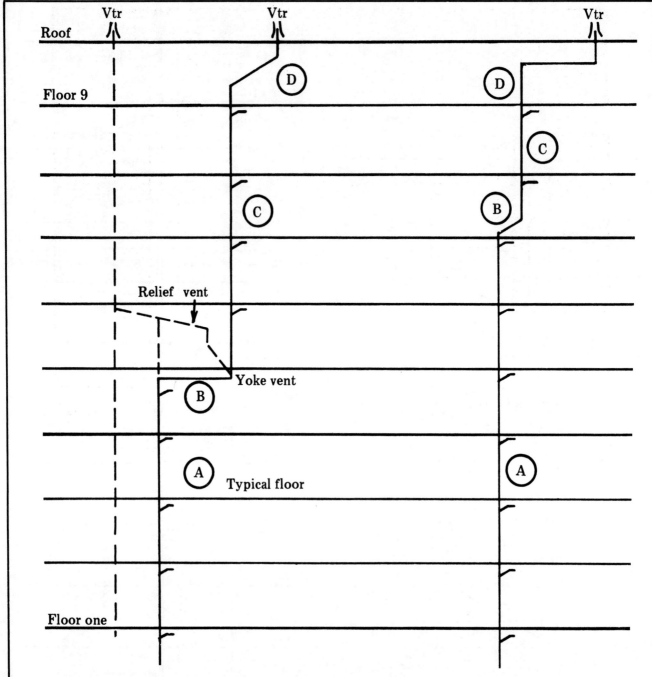

1. If offset of 90 degrees or less than 45 degrees as depicted in vertical stack Figure 8-33, the horizontal section of piping (B) must be sized as a building drain. If fixture units are great enough (A) and (B) would have to be increased in size while (C) would remain the same.

2. If offset is 45 degrees or less as depicted in vertical stack Figure 8-34, stack would be considered vertical throughout. No change in sizing piping in (A),(B), and (C) required.

3. Offsets of either 90 degrees or 45 degrees (D) in vent section would have no effect on pipe sizes in these figures.

Multi-story Plumbing Systems

Figure 8-33   **Figure 8-34**

Sanitary Sitework & Water Distribution Systems

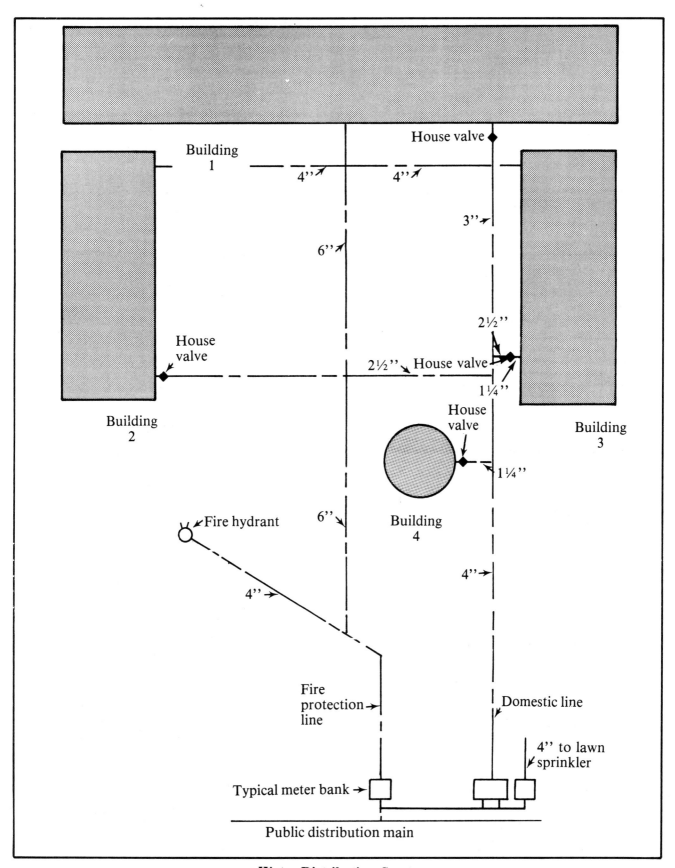

**Water Distribution System
Typical Building Site
Figure 8-35**

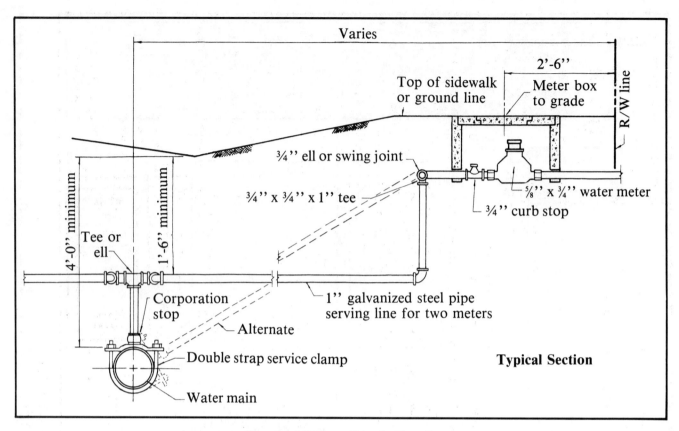

**Standard Water Supply Detail
Typical Service Installation for 4" or Larger mains
Figure 8-36**

Water Distribution Systems

For most small jobs the water sitework consists of a single water service pipe from the meter to the building. For sites having several buildings, a water distribution system is required. This will include the house meter and may include a fire meter and lawn sprinkler meter. See Figure 8-35. The sitework will include all the piping, house valves, fire hydrants and required meters, as well as the connection to the water main.

The plumbing contractor, in most instances, installs the water distribution system and may be responsible for tapping the water main and installing the required meters. Be sure to read this section of job specifications carefully.

Water Meter Installation

Public water mains may be located in a street, alley, or a dedicated easement adjacent to each parcel of privately owned property.

Upon request, the utility company will have their installation crews tap the water main and install the proper size meter. There is a specified cost, generally paid by the owner, general contractor or plumbing contractor. (Find out what this charge is when bidding the job.) If the water supplier won't do this installation, the plumbing contractor will have to make the connection.

The plumbing estimator should know the depth, size and location of the water main pipe. The municipal engineering department can supply this information. If sidewalk, curb or pavement must be removed to make this connection, add the cost of replacing these to the bid price.

A permit must be obtained from the local authority and an inspection will be required.

Figure 8-36 shows a typical water meter installation detail for a small job. Figure 8-37 is a standard meter bank detail similar to the one required for the building site plan illustrated in Figure 8-35. A typical valve setting that includes riser pipe and valve box is shown in Figure 8-38.

For large water distribution systems, air release valves should be installed to release trapped air in the water mains. A typical air release valve detail is shown in Figure 8-39.

Sanitary Sitework & Water Distribution Systems

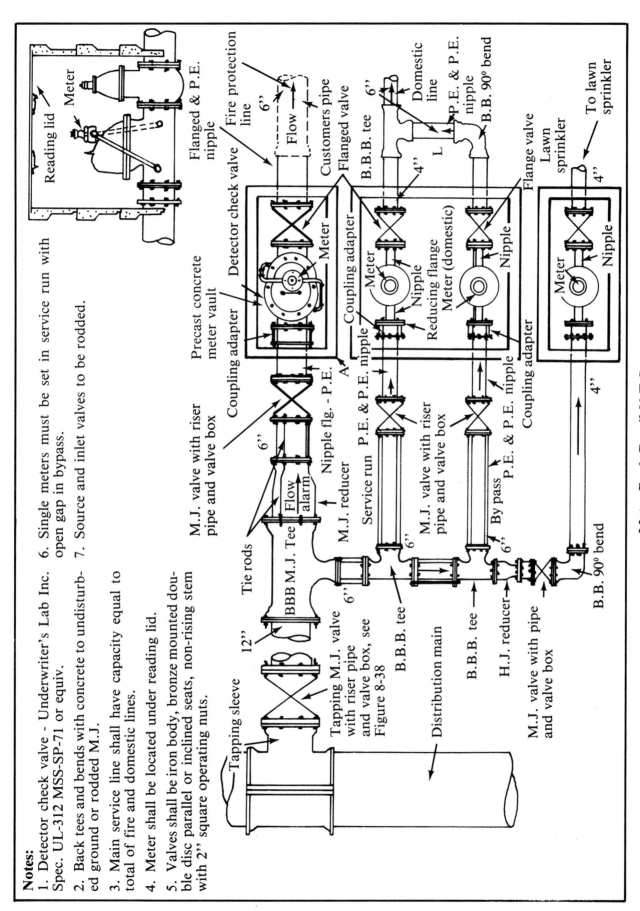

Meter Bank Detail N.T.S.
Figure 8-37

Notes:
1. Detector check valve - Underwriter's Lab Inc. Spec. UL-312 MSS-SP-71 or equiv.
2. Back tees and bends with concrete to undisturbed ground or rodded M.J.
3. Main service line shall have capacity equal to total of fire and domestic lines.
4. Meter shall be located under reading lid.
5. Valves shall be iron body, bronze mounted double disc parallel or inclined seats, non-rising stem with 2" square operating nuts.
6. Single meters must be set in service run with open gap in bypass.
7. Source and inlet valves to be rodded.

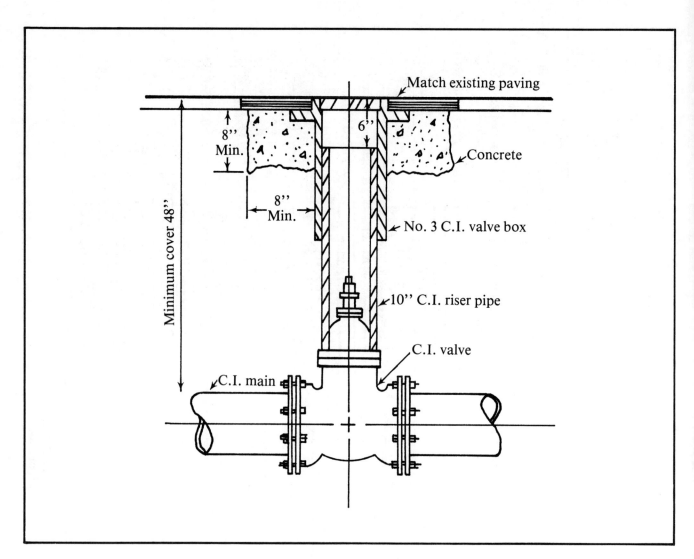

**Typical Valve Setting
(16" and Smaller)
Figure 8-38**

Sanitary Sitework & Water Distribution Systems

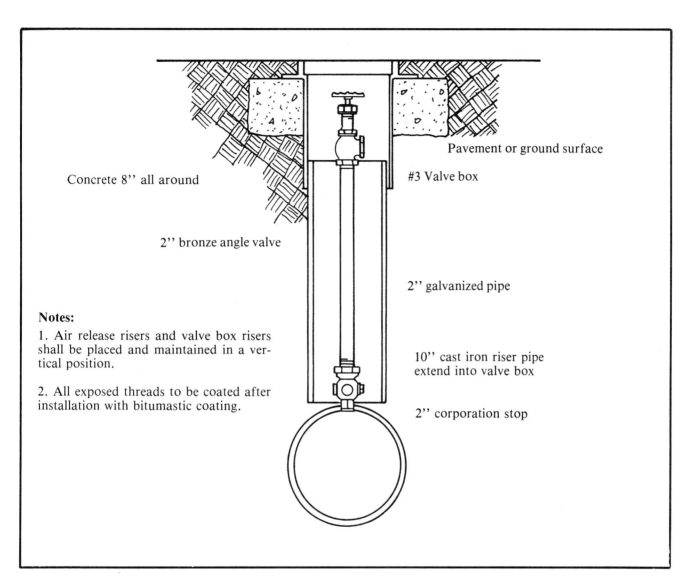

Typical Air Release Valve
Figure 8-39

Part 3

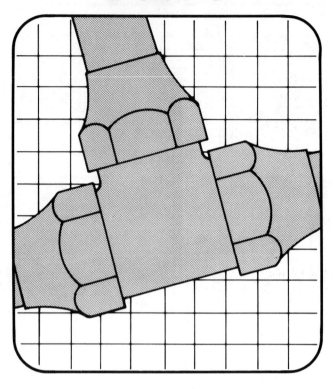

The Estimating Process

- **The Contractor & His Estimator**
- **The Plumbing Estimator**
- **The Subcontract Agreement**
- **The Material Take-off**
- **Estimating Forms**
- **Taking-off & Pricing The Estimate**
- **Completing The Estimate**
- **Man-hour Tables For Plumbing**

9

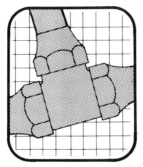

The Contractor & His Estimator

The construction cost estimator occupies a unique position in all industry. He must anticipate the cost of installing standard manufactured materials in remote locations where work is being done by other tradesmen under the direction of other contractors. And he can't change his price if costs exceed his estimate! No manufacturer would be willing to work under those conditions. But it's routine in the construction industry.

As an estimator you control both your firm's profit margin and the company workload. If you estimate for a small plumbing firm, you're probably the owner as well as the estimator. Even if you're not the owner, there's a lot you can and should be doing to help your company succeed. This chapter looks at several common mistakes that smaller plumbing contractors often make and outlines easy solutions that will help you avoid them.

Limitations

As a plumbing estimator you must know the size, complexity and quality of work your company should handle. Don't bid on work you don't understand. For example, many plumbing contractors have mechanics, tools and equipment well suited to single family residences, duplexes and small apartment units. But they don't have the equipment or experience needed for heavy commercial or industrial projects. It would usually be foolish to go outside your organization to get men with the technical ability needed to supervise such jobs. And it's expensive to rent or buy the special tools and equipment needed for such work.

Over-Expansion

Avoid over-expansion when economic conditions are good. The temptation is to take on more work than your firm can handle. Having too much work in progress can hurt as much as not having enough. Caution and self control will help avoid over-expansion.

The following are some of the negative effects of over-expansion:

• It can delay building progress schedules. Unnecessary delays are costly and loss of future bid opportunities could be the result.

• When the general economy is very good and construction employment is at a peak, productivity will be lower, thus raising the cost of the job. A firm expanding too fast will likely hire mechanics with marginal ability. This results in poor workmanship and additional "call backs" to correct inferior installations. Corrections are costly and produce dissatisfied customers. A bad reputation costs your firm more than the additional sales volume is worth.

Job Management

Every plumbing contractor is responsible for furnishing good job management. Good job management makes the difference between failure and completing a job on schedule and within budget. In a small firm, the plumbing contractor usually assumes this responsibility himself. In a larger shop, hired superintendents will be responsible.

The following are essential for good job management:

• Visit the job site and make necessary preparations *before* sending crews to begin the actual work.

• Install construction sheds or trailers if the job is to last longer than a week or two. Work benches, tools and materials should be conveniently located to avoid unnecessary travel from storage space to work areas.

• Select a qualified foreman who can work well with the men he supervises and with other tradesmen on the job.

• Furnish the job foreman with isometric drawings of the drainage, waste and vent systems. This is absolutely essential for larger buildings but is a big help on all work. You need isometrics to estimate the cost of the job. Let your foreman use the same drawing when he starts to rough it in. Isometric drawings are the means of communication between plumbing professionals. Take the trouble to have good drawings. Your job foreman should be able to read and interpret these isometric projections accurately.

• Furnish the job foreman with a complete set of approved construction blueprints and specifications, as well as roughing-in sheets for special plumbing fixtures or equipment.

• Make sure there are enough good quality tools for each crew to perform its installation properly.

• Delivery of materials should not interrupt the job. Delay is an unnecessary expense that can be avoided by proper scheduling.

• When safe, the roughing-in of underground piping should be done *before* the fill is added. This eliminates additional expenses for excavating and backfilling.

Know Your General Contractor
You have to estimate from plans and specifications which are prepared by architects and engineers. Your work crews must install plumbing materials in job and weather conditions over which you have little or no control. With so many unknowns in every job, you'll feel a lot better about your estimates if you know all you can about the general contractor you're bidding for. Some of the questions you should be concerned with before signing a contract include:

• Does the general contractor pay his subcontractors promptly? If he doesn't, it will cost your firm the cash discounts permitted by most suppliers. In a close bid, this delay can rob you of the profit in a job.

• Does the general contractor have a good reputation for efficiency and cooperation in dealing with his subcontractors? Slow construction progress can force you to work on another section of the same job or switch to another job. This could mean moving the crew and equipment and will increase labor costs and reduce profits.

10

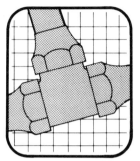

The Plumbing Estimator

Most plumbing contractors will agree that there's good money to be made in the plumbing business today. The opportunities are there for anyone who can assemble the personnel, the assets and the know-how to get the job done. But running a plumbing business is much more complex than many journeyman plumbers realize. Most of the larger plumbing contractors in your area probably got where they are by making all the mistakes and learning from each one. But that's a dangerous way to run any business. Mistakes like taking on too much work, stretching cash reserves too thin or underestimating costs can be fatal to any contracting business.

No book can explain how to set up a profitable plumbing contracting business. But most experienced plumbing contractors can tell you why so many plumbing contractors go broke or go back to working for someone else.

It isn't lack of work. There's seldom a time when good, profitable work isn't available. It isn't lack of good tradesmen or shortage of materials—this was a problem in the 1950's and occasionally since, but most plumbing contractors find ways to work around problems like these. It isn't high interest rates—a profitable, growing business can always get by, even when the cost of borrowed funds hits 20%. It isn't government regulations or OSHA or cutthroat competition or excessively high wages or union relations or most of the other problems plumbing contractors complain about when they get together. The real villain is poor estimating—to be specific, failure to estimate all the costs. And you can't blame anyone but the estimator for a poor estimate. If you don't know your true costs, only luck will keep your company in business.

Estimating material and labor costs is the single most critical part of any plumbing business. But there are other important costs too. Most of these fall under the title of overhead. That's what this chapter is about. The next few chapters look at labor and material costs in detail.

What Makes a Good Estimate?

Estimating is generally done by the master plumber who owns the firm in a small plumbing operation. His estimates either make his plumbing business a success or a failure. A good plumbing estimator usually has the following qualities:

• He must be a practical man with common sense.

• He must be familiar with the local plumbing code.

• He must have broad field experience in many types of jobs under various job conditions. This includes assembling materials and equipment on job sites, laying out the plumbing for installation, supervising men and coordinating work with the general contractor and other trades.

• He must be able to read and understand construction blueprints and prepare accurate isometric drawings.

• He should be able to visualize all the stages of plumbing installation.

• He should have a good knowledge of arithmetic, especially the decimal system. Practically all measurements and computations in a plumbing estimate are made in linear feet, square feet or yards, and cubic feet or yards.

Business Failures
Plumbing contractors fail for many reasons. Those that involve the plumbing estimator are:
• Failure to estimate the job completely. There is no substitute for a complete and detailed material take-off. Labor is based on material quantities. Overlook materials and you leave labor cost out of the bid. This can result in a ridiculously low bid and a major loss.
• Always being the lowest bidder. The habitual low bidder is wrong more often than right. If you get most of the jobs you bid on, you're bidding too low. Review your take-off method. Cross-check the figures on your work sheets. Mathematical errors can cost you plenty. A calculator with a tape makes accurate cross-checking easier.
• Shortcuts to save time. When business is booming, you'll have to estimate more jobs than you can handle. You might be inclined to take shortcuts in an effort to save time. This is a serious mistake. Shortcuts reduce your accuracy.

There are ways to save time without taking shortcuts. One is by using your experience from estimating similar jobs. Knowing what to estimate and how to estimate it always saves time. It ensures accuracy, too.

The Detailed Estimate
If several plumbing estimators were to estimate a particular job using the same plans and specifications, the spread between the high and low bidder might be as much as 20% or more. Why such a variation? There's no uniform method of estimating. Each plumbing contractor has his own system. But nearly all agree on one point: It's important to estimate in the order the materials will be installed. The next chapters take up detailed and systematic estimating procedures.

There are three distinct classes of work. Each should be estimated separately. First is the drainage, waste and vent system. Second is the water piping system. Third are the plumbing fixtures and equipment. Labor cost should be estimated separately from the material cost for each class of work. There is usually more variation in labor cost than in materials. Estimating labor separately allows you to compare the actual cost with the estimated cost for each class of work. This lets you see, for example, if you've underestimated labor for the installation of the water piping system or other classes of work. Separate labor estimates help you avoid similar mistakes in the future.

Office Overhead Expense
Most plumbing contractors are pretty good at estimating their labor and material costs. But many don't really know how big their overhead costs are—both job overhead and office overhead. All too often what looks like a good profit on a job is really a loss when office and job overhead are considered.

Office overhead includes all business costs not directly chargeable to any job. These must be paid regardless of the amount of work done. The size of your business determines these office overhead expenses: rent and utilities, office staff, vehicles, office insurance, advertising, legal and accounting fees, your salary as manager of the business, postage, office supplies, dues and subscriptions, travel and entertainment, health and retirement benefits for your office staff, and taxes. You can probably think of many other expenses associated with doing business. These are office overhead costs.

Overhead expense usually runs between 5 and 10% of a plumbing business' annual income. For most it averages about 6 or 7% of gross.

Most successful contractors keep their overhead as low as possible. It's a big advantage to keep the cost of running your office in the 6 to 7% range. As an example, company "A" has a fixed office expense of 6% of its estimated annual income of $1,000,000. Company "B" has expenses of 8% of the same income. "A" and "B" often compete for the same jobs. You can see that company "A" has a $20,000 advantage (2% of $1,000,000) over "B". "A" can lower its bid price on each job because of its savings on overhead.

There is no agreement on how office overhead expense should be included in your estimate. Some plumbing contractors figure that these expenses come to X dollars a week. If they have three jobs going and each is expected to last about a week, each job is charged 1/3 of X dollars as office overhead.

Other plumbing contractors use a different system. They assume that during the coming year they will have, for example, 10,000 productive man-hours on all jobs. If the office overhead is expected to be $75,000, $7.50 is added to the cost of each man-hour for estimating purposes. So if the cost of labor is figured at $20 per hour, $7.50 is added to cover office overhead.

There is another variation of this system. Suppose you know that your office overhead will be about $75,000 for the coming year and that you can expect to handle about a million dollars in work. Your office overhead is 7.5% of gross. Simply add 7.5% to your estimates. Of course, your costs can vary and your volume estimate may not be exactly accurate. That isn't the point. You may be off by a percent or two, but at least you didn't have a 100% miss.

"Cost Plus" Jobs

There are some jobs where fixed contract prices are not feasible. Proposals may call for compensation based on time and materials required. This type of contract requires you to state the percentage you will add to costs to cover overhead and profit. Overhead in this case would include both office and job overhead.

On cost plus work where cost and profit are guaranteed, you may be willing to take a smaller profit than with a bid contract. For example, three plumbing contractors estimating the same job may have the same combined fixed overhead expense of 10%. "A"'s bid is 10% plus 6% profit; "B"'s bid is 10 plus 5; and "C"'s bid is 10 plus 10. While the overhead remained the same for all three, the profit varies from 5 to 10%. Thus, the contract will be awarded to "B", the low bidder.

Job Overhead Expense

Many types of expense fall under the heading "Job Overhead." These are supervision (project superintendent, job foreman, etc.) construction sheds or trailers, prefabricating benches, tools, permits, sales taxes, insurance, bid bonds, warranties, and trash removal. Many of these job overhead items are significant costs and should be estimated item by item and listed separately on the estimate. Other job overhead costs are fairly small and can be assumed to be about the same percentage on all jobs. For example, you might figure that small tools and miscellaneous supplies are 1% or 2% of every job. Whether you itemize each of these costs or cover them with a single percentage of markup, these are actual costs just like labor and materials and have to be included in your bid. It isn't unusual for job overhead to come to more than 10% of the total cost, so be sure you include in your estimate every penny of job overhead that you can identify.

Most job overhead expenses can be estimated fairly accurately. Large plumbing contractors usually have their own accountants who follow these fixed expenses. This lets them determine where costs exceed estimates and where adjustments are needed on future jobs. Smaller plumbing contractors may not know where to adjust on future bids and may continue losing money until they fail.

This isn't a book on accounting, but in this area accounting and estimating go hand in hand. You've got to record your actual job overhead costs if you want to estimate future costs accurately. Fortunately, small computers have made it possible for even small contractors to keep good job cost records just like the larger contractors. But if you don't have a system of account codes and job files, no computerized job cost system will help. And don't think you need a computer to maintain good accounting records. Service bureaus all over the U.S. will help you maintain accounting records every month for a modest charge, typically under $50 a month.

It's poor practice to lump job overhead and office overhead into a single percentage and then add that total into every bid. Job overhead varies too much from job to job to be taken so lightly. It's much better to figure and list job overhead and office overhead separately. And try to break out and itemize as much of your job overhead as seems reasonable for the type of work you handle.

Notice that a part of your salary may show up both under job overhead (as the estimator or supervisor) and office overhead (as the manager and recipient of company supplied benefits). Don't assume that what you earn is chargeable only as office overhead.

Don't Forget These Job Expenses

Superintendent—Plumbing contractors handling several large jobs at a time need one or more superintendents. These men are considered non-productive overhead labor. Their salary and travel expenses should be estimated and charged directly to the job or jobs they oversee. These expenses are based on the estimated time the jobs are under construction.

Foremen—There are generally two classes of plumbing foremen: non-working and working. The non-working foreman is used on jobs requiring two or more work crews. His time is spent ordering and placing materials, laying out each day's work, organizing the work crews, and keeping records for the payroll department. His salary should be charged directly to the job for the estimated time the job is under construction.

A working foreman is generally assigned to supervise one work crew, for example, the water piping crew. Supervising will not consume all of his time. A certain percentage of his time is productive labor. This percentage is difficult to estimate. Generally a working foreman's time is equally divided between non-productive and productive labor. His labor cost is estimated from the time the job is under construction. One-half of his payroll cost should be charged as a direct job expense.

Construction shed or trailers—The rental or cost (prorated over the shed or trailer's estimated useful life) should be charged as a direct job expense for the estimated time the job is under construction.

Temporary toilets—Chemical or portable toilets must meet local code requirements. These sanitary facilities are usually provided and paid for by the general contractor for all workers. In some cases toilet facilities requiring water and waste disposal may be required. These would be installed by the plumbing contractor and included in the job contract. This cost should be charged as a direct job expense.

Temporary water—On new construction, water for construction purposes is usually provided by the plumbing contractor. On single-family residences or other small jobs, this may consist of a short run of pipe with faucets for hose connections. High-rise buildings require water outlets on each floor. Labor and material for this installation should be charged as a direct job expense.

Sleeving and cutting—On large jobs the plumbing contractor is expected to sleeve pipes that pass through poured concrete. The openings left between the pipes and the concrete must be properly patched to meet code requirements. General contractors usually assume this responsibility. However, the plumbing contractor's bid should make it clear that all patching around plumbing pipes is assumed to be done by others.

Trenching and backfilling—The plumbing contractor should trench and install all underground pipes before finish grading is completed. This includes interior as well as site work. The cost of trenching is charged directly as a job expense. The plumbing bid should make it clear that backfilling is to be done by others.

Plumbing permits—Plumbing permit prices vary from city to city. A set charge is made for each plumbing fixture, appliance and appurtenance, as well as for water service, sewer connection, roof drains, etc.

Building and zoning departments for each city have a fee schedule. The cost of plumbing permits should be charged as a direct job expense.

Trash removal—There's always trash to be removed from every building site. Avoid the possibility of being back-charged by the general contractor for removing the plumbing portion of the trash. Note in your bid that removal of trash is to be done by others.

Sales taxes—Many states have sales taxes. This tax usually must be paid at the time materials are purchased from the supplier. You have to add sales tax to your estimate. Sales tax is another job expense.

11

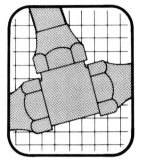

The Subcontract Agreement

Your contract with the general contractor is an important consideration when estimating any job. The contract spells out your responsibilities and the general contractor's or owner's duties. Everything in the contract can affect job cost, so you should review the documents you have to sign before finishing the estimate.

Most of all, you want to avoid hassles and surprises that can run costs way up. If you can see problems coming, include a "grievance" allowance in your bid. Better yet, practice a little preventative law by resolving potential disputes before they arise. Use contract documents you know and understand if you have a choice. And most of all, put your agreements in writing, even for small jobs if they are to be performed over a period of time longer than a few days or if there are several parties to the agreement. Don't rely on "side agreements" or oral understandings that modify a written agreement. Don't agree to do something you may not be able to do or pay a sum you may not have available.

Don't contract with anyone until you are sure that they have the ability and desire to pay their bills on time. It is a fact of life in the construction industry that many "sharp" general contractors and speculators do not pay on time. They realize that even if you have a 100% valid claim against a solvent debtor, if you have to use an attorney to collect, you give up at least half the amount owed. The attorney's fee will be between 1/3 and 1/2 of the debt and you will probably waste between 1 and 3 years getting a judgment against your debtor.

Somewhere during that time you will become anxious to settle for something less than the full amount of the debt. The result: you settle for 75 cents on the dollar and your attorney takes about half of that. Your "sharp" operator gets a 25% discount and delays payment several months at least. Meanwhile you have to pay your craftsmen and material suppliers.

Avoid disputes by avoiding surprises. No one is surprised if everyone does what is expected. The building contract, whether written or oral, outlines what is expected of the parties contracting. The plans and specifications are a part of that contract and impose an obligation on the builder and on your plumbing firm. These obligations should be as clear and precise as practical and should anticipate as many problems as can reasonably be foreseen. Don't experiment with drafting your own contract and don't sign a contract offered by others if you can get them to accept your "standard" contract. Some contracts favor subcontractors and some don't. Local and national associations offer "standard" contracts that favor and protect the subcontractor. Have the contract drawn on one of these documents if possible. If this isn't possible, read the contract the general contractor insists on. You are held to everything in the contract you sign even if you don't read it. If something seems unfair to you, offer to cross out that portion on all copies of the contract. If the other parties claim that the offensive portion of the contract isn't relevant, you should reply that it is better then to eliminate it by striking it out. If you

Proposal

FROM

Proposal No.

Sheet No.

Date

Proposal Submitted To	Work To Be Performed At
Name	Street
Street	City_____ State
City	Date of Plans
State	Architect
Telephone Number	

We hereby propose to furnish all the materials and perform all the labor necessary for the completion of
_____
_____
_____
_____
_____
_____
_____
_____
_____

All material is guaranteed to be as specified, and the above work to be performed in accordance with the drawings and specifications submitted for above work and completed in a substantial workmanlike manner for the sum of _____ Dollars ($_____). with payments to be made as follows:
_____
_____

Any alteration or deviation from above specifications involving extra costs, will be executed only upon written orders, and will become an extra charge over and above the estimate. All agreements contingent upon strikes, accidents or delays beyond our control. Owner to carry fire, tornado and other necessary insurance upon above work. Workmen's Compensation and Public Liability Insurance on above work to be taken out by _____

Respectfully submitted _____

Per _____

Note — This proposal may be withdrawn by us if not accepted within _____ days

ACCEPTANCE OF PROPOSAL

The above prices, specifications and conditions are satisfactory and are hereby accepted. You are authorized to do the work as specified. Payment will be made as outlined above.

Accepted _____ Signature _____

Date _____ Signature _____

Subcontract Agreement or Proposal
Figure 11-1

The Subcontract Agreement

want something in a contract that isn't there, write it in on all copies and have the parties initial the addition. The addition becomes a perfectly legal part of the agreement. If you don't understand your obligations, get the advice of an attorney or pass up the job completely. In any event, don't start the job without a written contract. The Statute of Frauds in most states won't let you collect on many oral contracts. At best you will have to convince a court that there was really a contract; at worst you will only be allowed to collect the reasonable value of your services rather than the full contract price. In either event you will have to go to court to collect what you are owed.

Elements of a Contract

Courts will not enforce an agreement unless certain traditional elements can be found. For example, "lawful consideration" is essential to a contract: There must be an obligation on both parties to do work, deliver money or property or give up a legal right. A moral obligation isn't enough. Also there must be a clear, flat, unconditional acceptance of an offer and the acceptance must be communicated. The element of mutuality must be present: an obligation upon both the subcontractor to do the work and upon the general contractor to pay for the work. A simple memorandum of prices, even though signed by the parties, isn't a binding contract in the absence of an understanding by the parties to perform the agreement.

The contract must be so worded as to allow the intent of the parties to be understood with a reasonable degree of certainty. If the specifications are so indefinite and plans so poorly drawn that it's impossible to interpret them, the contract will not be enforced by the courts.

Compensation

As the work progresses, you should be entitled to partial payments. The amount of the partial payment should consider the amount of material delivered to the site and work done at your shop as well as completed work.

Partial payments usually cover from 90 to 95 percent of the work completed during the payment period, 10 or 5 percent being retained as protection against claims or charges resulting from some act of omission on your part. Possible grounds for such claims or charges include:

- Your failure to pay material dealers or employees.
- Damage to adjacent property or to another contractor.
- Defective work not corrected.
- Evidence that the work cannot be completed for the unpaid balance.
- Evidence that the contractor may default.

When final payment is made, you may be required to deliver a release of all claims or liens arising out of the contract. If any claims or liens are outstanding, the amount claimed may be deducted from the final payment and retained until the matter is settled. If you have any unsatisfied claims for damages, extra work, or other reasons, releases should exclude the unsatisfied claims.

Penalties and Indemnity Clauses

The matter of penalties for non-performance by a specified time is one that requires careful handling. The courts are uniformly against the enforcement of clauses providing for penalties, forfeitures and even liquidated damages. If you can show that the owner or general contractor did not suffer any damages, or that the damages suffered are materially less than stipulated in the contract as liquidated damages, the court is likely to refuse to enforce such a clause and grant the owner only the damages that were actually suffered. Liquidated damages provide a specified forfeiture for a specific failure and will usually be enforced if it appears to be a good estimate by the parties of what the actual loss is and the exact loss is difficult to calculate. For example, many contracts require completion by a given date. Each day completion is delayed beyond that date results in a penalty to the contractor of a certain amount. If the penalty seems reasonable, the court will probably assess damages accordingly, especially where there is some loss and the actual damages are difficult to calculate.

"Cost Plus" Contracts

If you undertake work on a time and material basis, realize that some courts have ruled that the contractor could not charge any overhead on the job. Of course, particular words or clauses in the contract might provide reimbursement for job and office overhead. Be very careful in "cost plus" contracts as to exactly what costs are included. Overhead is a very real cost. Itemize all the costs you intend to be paid for.

Cancellations of Contract

Contracts frequently provide that in the event of delay or certain defaults, the owner may cancel the contract and complete the work himself at the expense of the builder. Generally these clauses are made contingent upon a certificate or report of an architect or other agent of the owner, that a contractor has unreasonably delayed the work or refused to comply with the contract terms. While it seems unfair that an architect hired by the owner should have such authority, the right to cancel can arise only after some act that is clear and definite. If the architect acts unfairly or fraudulently, you have a remedy in court to show fraud or collusion.

Completion Problems
It is both unfair and unrealistic for a contract to require completion by a certain date without regard to circumstances beyond your control. Your contract should permit delays for strikes, poor weather, unavoidable material shortages and many other events that can prevent completion. If the contract specifies that "time is of the essence," a delay, even unavoidable, may mean that you have breached the contract and cannot recover your profit. Make the contract clear on this issue and don't do business with anyone who insists on unrealistic completion dates in the face of insurmountable problems.

Courts don't recognize your right to stop work if the job begins to look unprofitable. But there are ways to protect yourself. First, try to eliminate uncertainty by specifically excluding conditions over which you have no control or about which you have no knowledge. Exclude from your bid anything that is impossible for you to estimate. Write up the agreement to provide additional compensation if difficult conditions are encountered or if material or labor costs go up. If you find an error in your bid after it has been accepted, go to the owner or contractor and ask to be relieved of the obligation to complete the job as bid. There are many situations where contractors have been excused from performance after an erroneous bid. The bigger the error, the more the owner or general contractor should have realized that an error was made. Accepting a bid in bad faith with knowledge that it is based on an error is grounds for cancelling the contract.

Every bid is not an offer to form a contract. Your estimate form could contain language to the effect that the estimate "is a bid only and not an offer to perform the work at the stated price." If the owner or general contractor approves the estimate, you may want to go over it once more very carefully to be sure that everything was included and the job will yield a reasonable profit. Any written offer you make should be dated and have a time limit for acceptance. A day or two is enough in many cases. Otherwise, an owner or general contractor may wait several weeks or months before agreeing to go ahead.

Breach of Contract
Under contract law if one of two parties to a contract fails to perform an important part of the contract, that party has *breached* the contract. The other party is excused from further performance and is entitled to recover his loss. Occasionally you may have to decide if an owner or your general contractor has breached the contract and you are justified in pulling your crew off the job and suing to collect your loss of profit. If you don't stop work, you waive the right to claim a breach of the contract and must continue to perform. If work stops and a court decides there was no breach of the contract, you may have breached the contract by stopping work. In most every case the failure to make a progress payment is a material breach of contract and relieves you of the obligation to continue. A project behind schedule is not a breach of contract unless "time is of the essence" of the contract and perhaps not even then unless the delay is serious. A flat statement by either party that he has no intention of completing the contract is usually considered an "anticipatory breach" and relieves the other party of the obligation to go further.

If you face a situation where you believe the opposite party has breached the contract but you have some doubt, it is wise to get legal advice before doing anything that might place you in breach. Then, armed with the advice of counsel, go to the owner and explain to him the consequences of continued failure to perform. You may convince him to continue under the contract. You will probably drive him to seek counsel of his own (who may convince him to go ahead under the agreement). And you will almost certainly prevent yourself from falling into a breach of contract by non-performance.

Follow the Contract and Specifications
Follow closely your specifications and you will not be held responsible for any part of the work found to be improperly planned by other parties. Where there is a conflict between the written specifications or contract and the drawings, the written provisions will always govern. If it's necessary to make changes in the contract or specifications, or if there are vital omissions, let these be cared for by written instructions from someone having full authority. Don't take things for granted; don't assume that anything will "be made all right" later on; get it in writing. Most contracts provide that the contractor will be reimbursed for the cost of changes or "extras." This sounds good but really is not enough. Changes are costly and time-consuming and any plumbing contractor who accepts changes during construction is entitled to considerably more than

his out of pocket cost. Overhead and profit should be included in the cost of changes just as they are in the basic estimate. Changes should be made only at your selling price.

Understanding Your Contract
The law works on the assumption that if you sign a contract, you know the contents of it and the obligations you are assuming. If you don't know the contents of the contract or understand all the provisions, you may find it a costly contract to carry out. Read the fine print!

If you draw up the contract, don't provide that the work shall be done to the satisfaction of the owner, leaving him as sole judge. Any other phrase that implies a good workmanlike job is far more acceptable.

One contract you are likely to see is that of the American Institute of Architects. This is a well-prepared, fair document. Read a copy and digest its conditions, paying particular attention to articles covering protection for owners on liabilities, authority of the architect, stopping work, procedure and pay for changes, breach of contract, arbitration, contingent responsibility, applications for payments and architect's certificates.

Faulty Work
In considering the question of faulty work, the courts are inclined to examine the matter of intent and good faith. For this reason decisions appear to contradict one another when examined hastily. A plumbing contractor used soil pipe made of inferior materials. The defective condition could not have been discovered by careful inspection, but developed only after exposure to the soil. If an architect had the authority to construe and determine the manning of the specifications and plans, and had the power to accept or reject materials and workmanship, and the structure has been accepted and the contract fully executed and no fraud has been involved, it is well settled that the owner will have no recovery for defective work or materials.

Extra Charges
Many of the differences that arise between the subcontractor, general contractor and the owner develop from the question of "extras." This is natural as the owner probably feels that the cost is being increased unnecessarily and without reason. You probably feel that you are being called upon to do work not included in the contract and for which extra compensation is due. Any changes or extras should be done only upon authorization in writing by the owner or his representative. Should the owner directly authorize extras, there is no question as to his liability if the work is done according to instructions. If the architect authorized the work, it brings up the question of whether he was the agent of the owner and his right to bind that owner for the additional work.

The terms of your building contract will or should help you on who can authorize extra charges. The status of the architect and his authority should be stated clearly in the contract. If no authority exists in the contract, the owner can refuse to accept the charge, leaving you to try to recover from the architect personally. Remember, the architect's powers are limited by the terms and conditions of the agreement between the architect and the owner, and you should be informed of the scope of their agreement.

Investigate All Conditions
Unexpected conditions which result in expenses not previously provided for in the estimate do not necessarily justify an additional charge. The owner can deny such a claim because he assumed you investigated conditions and entered into the contract with knowledge of the work required. Of course, if it can be shown that the added costs are the result of an act of omission on the part of the owner and not your fault, you will recover for completing the contract under the altered conditions. In general, if you meet conditions which make it impossible for you or anyone to complete the contract at any cost, you will be relieved of the obligation to finish the job. But just finding extremely hard rock where a trench should be will not be a solid basis for claiming additional compensation. It is better to exclude from your bid all conditions over which you have no control or which you have not investigated.

Substantial Performance
The idea of substantial performance was developed by the courts to do away with the unfairness of an absolutely literal performance of the contract and the hardship it would work on you. If you comply with each and every one of the important elements and provisions of the contract, you can demand reimbursement. Conditions determine in each case what constitutes substantial performance. In general, courts will not require you to perform an idle act which is of no benefit—even if the contract specifically requires it. For example, an owner cannot withhold payment on a job because the pipe joint compound was not precisely as specified if the quality of the joints is equal to or better than that specified. If your performance is only somewhat less than required, you will not recover the full con-

tract price and may have to reimburse the owner for any loss he suffers as a result. If your work does not meet the test of substantial performance you may not be able to recover anything or only the value of the incomplete structure. For example, you contract to plumb a home. Upon completion it is pointed out that the sewer drainage system is faulty and the home cannot be approved for occupancy by the local building authority. You have not substantially performed the work required. The structure has little or no value to anyone and courts most likely will deny any recovery. Note that "substantial performance" may be indicated by the owner. If the owner goes into occupancy of the structure and uses it for its intended purpose, it will be difficult for him to deny that the work was substantially performed.

The Relation of the General and Subcontractor

Under many building contracts you are the general contractor's subcontractor and are not recognized by the owner. The right to approve or disapprove subcontractors may be reserved by the owner in the prime contract. In other matters, the owner will probably hold the general contractor directly responsible for your work. It should be noted, however, that the subcontractor has definite legal rights which have important implications to the owner. For example, if the general contractor should default on payments due you, the owner will probably be held responsible for the debt and payment could be enforced by a mechanic's lien on the structure. The owner should require the general contractor to furnish releases from all claims in connection with the work, executed by all subcontractors, before final payment is made to the general contractor.

Mechanics Lien Laws

A lien is a claim created by law upon buildings and other improvements for the purpose of securing a priority of payment for value of work performed and materials furnished in erecting or repairing a building. The lien attaches to the land as well as the building.

Lien laws are in force throughout the United States and Canada. The general principle upon which lien laws are based is that any person who has contributed his labor or furnished material for a structure erected by an owner upon his premises should have a claim upon the property for his compensation. It is applied to the mechanic, workman, material man and others contributing direct labor and materials.

Whatever the nature of the contract, if substantial performance can be proven, a lien may be issued and enforced. Unimportant omissions on the contract terms will not forfeit the lien rights.

A delay in carrying out a contract will not prevent you from enforcing your lien rights if the contract has actually been carried out. However, the owner or general contractor may set up a claim and receive reimbursement for damages if proven. Naturally, if the contract work is based upon a stipulated time element, a serious delay may be a breach of contract that prevents enforcement of lien rights.

In taking advantage of lien privileges, you must abide by your contract in its essential requirements. This may include securing the architect's approval of work as evidence of payments earned.

Many states give subcontractors a priority lien on the amount of money due the general contractor at the time lien issues. If the contractor fails to complete and the subcontractor makes a new contract with the builder, some states allow the subcontractor lien on the full amount of value of such work as the original contractor abandoned. In some states the ruling is that no lien can attach unless the contract is recorded.

Some states give those who have performed labor and have furnished materials priority over creditors of a general nature of an insolvent owner or builder. These same states provide for the priority of day laborers over contractors and subcontractors, regardless of when their liens are filed; likewise a material man has priority over such contractors or subcontractors.

12

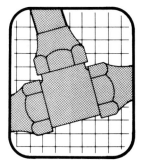

The Material Take-off

Every plumbing estimate begins with an accurate, detailed material take-off. Your aim is to include every material cost item on the job being estimated. Think of your take-off as the foundation on which your selling price is built. The material quantities are the basis for labor estimates, equipment estimates and overhead estimates. The reasons for accuracy in estimating materials should be obvious.

You'll have no trouble preparing accurate, detailed estimates if you have a good understanding of the plumbing code and know how plumbing is done. But you will have to study the construction blueprints and structural details carefully. And you have to understand the job specifications.

As you study the plans and specifications, occasionally you will find a conflict between the two. Remember, between the blueprints and specifications, *the specifications take precedence*. The engineer or architect should make the necessary change through a written addendum if you find a conflict. If a conflict exists between the blueprints, specifications and the applicable plumbing code, *the code is the final word*. Any conflict with the code should be brought to the attention of the job engineer or architect.

The Specifications
The specifications can be complex, consisting of a bound book with many pages of detailed information. Or they can be simple, consisting of a single sheet on the same page with the blueprints. This depends on the size, complexity, and type of construction.

Begin by reading the specifications carefully. List and make notes on each cost item in the job specs. Items which are expensive and easily overlooked include: water meters, the type and weight of materials, special fixtures, special trim, the type and make of valves, and whether bathroom accessories such as partitions and towel dispensers are included as plumbing work. These notes become a working index that is your reminder to account for these costs. This helps reduce costly omissions and errors.

The Blueprints
Give the same careful consideration to the drawings and structural details. First, be sure the blueprints are complete and that they cover all areas requiring plumbing. Note all the areas in the structural details that require offsets around beams, as well as partitions that do not line up one above the other in a multi-story job. Offsetting piping or hanging piping from a ceiling can be expensive.

Next, review the plot plan. It shows the source of water, the sewage disposal line, site drainage, gas lines, fire lines, and other essential information. Single-family and apartment plot plans are generally drawn to a scale of 1 inch equals 40 feet (1'' = 40'). See the residential plot plan in Chapter One, Figures 1-1 through 1-3.

Architectural sheets include the floor plans. From these pages you determine the type and quantity of plumbing fixtures, appliances, and equipment. Single-family and apartment floor plans are generally drawn to a scale of 1/4 inch equals 1 foot (1/4" = 1'). Commercial and industrial floor plans are generally drawn to a scale of 1/8 inch equals 1 foot (1/8" = 1'), or 1/2 inch equals 1 foot (1/2" = 1') for areas requiring toilet facilities.

Structural sheets provide additional information. This includes the type and size of building foundation, location of columns and beams, height between floors, and roof type.

General Estimating Outline
After reviewing the specs and plans, begin your actual listing of materials and quantities. It's important to follow a consistent procedure when doing the take-off. Consistency avoids errors. A suggested outline of the estimating sequence follows. Follow this or some similar outline throughout your estimating career. The outline divides the components of a plumbing system into logical sections and classes of work. This is your reminder that each section and class of work must be accounted for as part of the total cost. The take-off sequence outline is shown below. Chapter Fourteen shows in detail how to use this outline and includes a sample estimate.

1. Site utilities
2. Sanitary drainage, waste and vent system
3. Storm drainage system
4. Hot and cold water system
5. Fire protection system
6. Special piping system
7. Swimming pool
8. Temporary facilities
9. Plumbing fixtures
10. Accessories
11. Cleaning fixtures (if not excluded from the contract)
12. Final inspection

Using Man-Hour Tables
Once all material quantities are listed, you can begin the labor estimate. The man-hour tables in this manual will be a good guide to the time required for installing each component in a plumbing system. For example, you will list the number of residential tank type floor-mounted water closets on your fixture take-off sheet. Multiply this number of fixtures by the time required for installation of each. Multiply that total by your hourly labor cost to determine the total labor cost for this task. This procedure is used to figure the installation cost of all items from the material take-off sheets.

Here's how you would figure the installation time on a residential tank type floor-mounted water closet. Most plumbers should be able to make a simple, close coupled water closet installation to an existing waste and water outlet in 1 hour or less. Should you allow just 1 hour's installation time in the labor take-off? The answer is an emphatic "no." There are several additional costs to consider. You must be as precise as possible yet protect your firm against losses. Can you afford to "guestimate" an extra 1 to 3 hours to cover these additional costs? Probably not. But there *is* a way to analyze the extra time that is nearly always required on plumbing work.

Certain required and less obvious items are likely to increase actual installation time for the bowl and tank. Times for these items are included in the man-hour production tables in Chapter Sixteen. Some of these additional time requirements are:
• Handling at the job site.
• Grouting the bowl base where it contacts the floor.
• Installing the water closet seat.
• Testing and adjusting the water closet for final inspection and acceptance.

Each of these standard procedures should take no more than 15 minutes. This brings the total time needed for water closet installation to nearly two hours.

Is two hours an adequate time allowance? No, for time must be allowed to cover mistakes and problems that are common on many if not most jobs. For example:
• The plumber may have mistakenly roughed in the waste outlet too close or too far from the wall. This should happen very seldom. But when it does, considerable time is required to make the correction.
• The wrong color or type of water closet may have been shipped to the job site. This requires extra time to make the exchange.
• Sometimes damaged or defective water closets arrive at the job, requiring more time for correction.

Anyone who has worked on a plumbing job can identify many other reasons why production may be less than normal. Allow an additional 1/2 hour of labor per water closet to cover these problems and errors unless you are the rare plumbing contractor who employs only tradesmen who never make mistakes.

Extensive man-hour production tables are in Chapter Sixteen. A portion of one of these tables is reproduced as Table 12-1 following. The figures shown reflect realistic installation times for water

The Material Take-off

closets.

Multiply the number of water closets by the man-hours by the hourly labor cost. For example, if 2 flush valve floor-mounted water closets are to be installed, the estimator would multiply 2 times 2½ hours, times $20.00 (hourly labor cost). This produces an actual labor cost of $100.00 to install the 2 water closets.

Use this procedure to estimate each class of work in a plumbing system. It should yield realistic yet competitive estimates.

Item Water Closet	Man-Hours For Each
Flush valve floor-mounted	2.30
Flush valve wall-mounted	3.00
Tank type, floor-mounted	2.30
Tank type, wall-mounted	3.10
Tank type, tank wall-mounted	3.15

Table 12-1

13

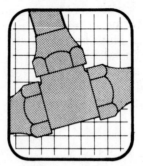

Estimating Forms

Quantity Cost Sheets

The foundation for every plumbing estimate is the quantity of plumbing fixtures to be installed, the amount of equipment required for the installation, and number, size and type of piping material. The time required to install each class of work varies with the material quantity.

List each work item separately as though each were a separate and distinct job. This makes it easier to check and price the materials and labor on the quantity sheets.

Before starting the actual take-off, write the following information at the top of your quantity sheets:

Job Address _____
Sheet Numbers _____
Date _____
Estimator _____
Checked By _____
System Type _____
Material Type _____
Title of Quantity Cost Sheet _____

The material and labor take-off forms should reflect your estimating needs. The forms in this book will be good for many plumbing contractors. Use them until you develop forms more suitable for the type of work you do. A properly designed material and labor take-off form will save hours of valuable time.

Standard take-off forms are available, but they may not be suitable for your work. Many small plumbing contractors prefer to use a *standard analysis sheet*, as illustrated in Figure 13-1. Be sure to identify each sheet properly with the basic job information to avoid omissions and errors. If standard take-off forms are not satisfactory for your needs, design your own estimate forms.

The take-off forms used for the sample estimate in Chapter Fifteen have been designed and used extensively by this author. They are simple in design, easy to use and have the virtue of keeping material and labor on separate work sheets for each class of work.

The Quantity Cost Sheet in Figure 13-2 is used for listing quantities of general roughing-in materials, the linear feet of piping, pipe fittings, valves, hangers and miscellaneous items. Each class of work is taken off, entered and priced under a separate heading. Examples of quantity and cost sheets are shown in Figures 14-5 through 14-11 of Chapter Fourteen.

The Quantity Cost Sheet in Figure 13-3 is used for listing quantities of plumbing fixtures or equipment. Ample spacing between lines allows room to identify the fixture, equipment type and trim schedule. Each fixture or piece of equipment is taken off the plans, entered under the proper heading, then priced. Figure 14-10 in Chapter Fourteen for an illustration listing fixture quantities. Equipment is not included in the sample estimate. Figure 14-11 is an example only.

J. Smith, 602 N.E. 89th Street

Drainage, waste and vent system

No-hub cast-iron — piping quantity cost sheet

	Initials	Date
Prepared By		
Approved By		

Sheet No.____of ____

	Description	1 Number	2 Unit Price	3	4 Cost	
1	4 x 10' cast-iron pipe	48 L.F.	2.21 L.F.		106 08	1
2	3 x 10' cast-iron pipe	112 L.F.	1.53 L.F.		171 36	2
3	2 x 10' cast-iron pipe	34 L.F.	1.27 L.F.		43 18	3
4						4
5						5
6						6
7						7
8						8
9						9
10						10
11						11
12						12
13						13
14						14
15						15
16						16
17						17
18						18
19						19
20						20
21						21
22						22
23						23
24						24
25						25
26						26
27						27
28						28
29						29
30						30
31						31
32						32
33						33
34						34
35						35
36						36
37						37
38						38
39			List Price		320 62	39
40			Discount 10 + 10		-60 92	40
			Net cost		259 70	

Standard Analysis Sheet
Figure 13-1

Center Plumbing Corp.

418 S.W. 16th Street

Hollywood, Florida 33021

(350) 579-1546

Job Address _____

Sheet Number _____ Date _____

Estimator _____ Checked By _____

System Type _____

Material Type _____

Quantity Cost Sheet

Description	Number	Unit Price	Cost
			List Price
			Discount
			Net cost

Figure 13-2

Estimating Forms 129

| Center Plumbing Corp.
418 S.W. 16th Street
Hollywood, Florida 33021
(350) 579-1546 | Job Address _____
Sheet No. _____ Date _____
Estimator _____
Checked By _____ |

Quantity Cost Sheet

Description	Number	Unit Price	Cost
		Net Cost	

Figure 13-3

Center Plumbing Corp.

418 S.W. 16th Street

Hollywood, Florida 33021

(350) 579-1546

Job Address _____

Sheet No. _____ Date _____

Estimator _____ Checked By _____

System Type _____

Labor Cost Summary Sheet

Description	Number	M.H.	Total M.H.	Rate	Cost

Total Man-hour Cost

Figure 13-4

Estimating Forms

Center Plumbing Corp.

418 S.W. 16th Street

Hollywood, Florida 33021

(350) 579-1546)

Job Address _____

Sheet No. _____ Date _____

Estimator _____

Checked By _____

Project Summary Sheet

Item	Classification of Work	Material	Labor	Total

Total Job Cost

Materials and labor costs are transferred to this sheet. Be sure to add all supplementary costs, as described in Chapter Fifteen.
Figure 13-5

Labor Cost Sheets

The Labor Cost Summary Sheet (Figure 13-4) is used to list the man-hours required to perform each individual operation. Total man-hours times the hourly labor cost, provides the total labor cost for each class of work. Labor for all classes of work should be entered on the labor cost summary sheet under the appropriate heading. Sample labor cost summary sheets listing man-hours and labor cost for each class of work are shown in Chapter Fourteen, Figures 14-12 through 14-14.

Project Summary Sheet

The estimate is now almost complete. The last step is to total each estimate sheet and carry the amounts forward to the Project Summary Sheet. A suggested form is shown in Figure 13-5. An actual sample of one is provided in Figure 15-1 as a part of the total Sample Estimate shown in Chapter Fourteen.

The Project Summary Sheet should include your overhead and profit, all permits, sales tax, insurance, bid bonds and other job costs. Total all costs to arrive at the bid price.

Quantity cost sheets, labor cost summary sheets, and project summary sheets will be discussed further and illustrated in Chapters Fourteen and Fifteen.

14

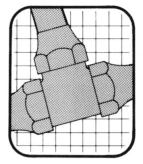

Taking-off & Pricing The Estimate

Assume you have to submit a bid for plumbing a single family residence. This chapter illustrates the procedure for taking off materials and figuring the labor cost. The procedure would be the same for any size or type of building.

The sample estimate pages in this chapter show the material quantities taken from the plans and the cost of those materials. Labor costs and material prices vary considerably from area to area and change over time. The labor costs and material prices used in the sample estimate are for illustration only.

Water piping diagrams and sanitary isometric projections are generally not included with the plans or required for small jobs. Anything prepared by architects for larger jobs will generally meet code requirements. But you should remember that any supplied layout may not necessarily meet job conditions. Don't use it for your material take-off. Instead, estimate the job the way you will actually do the work.

Begin the take-off by drawing an isometric of the DWV system. (See Figure 14-1.) Also, draw a piping diagram for the hot and cold water system. Refer to Figure 14-2. These designs should reflect your judgment of job conditions and meet prevailing code requirements. Next, size all piping for the two systems. This establishes the individual fitting sizes within the DWV and water piping system.

Select the type of code approved roughing-in material you will use. This may be specified by the architect. In the sample estimate, no-hub service weight (hubless) piping and fittings will be used in the DWV system and drainage copper piping and fittings for fixture branches (fixture drains in some codes). Type L copper piping with wrought copper fittings will be used for the hot and cold water system.

Use an architectural scale rule for taking off the linear feet of piping from the plans. Use a rule with 4 edges and a total of 8 scales. The scale the architect used in drawing up the plans is noted at the bottom of each sheet. Be sure the scale you use is the same as the one noted on the plans. Consider this mistake: If you use a scale of 1/4 inch equals 1 foot (1/4 = 1'), and the blueprints were drawn to a scale of 1/8 inch equals 1 foot (1/8" = 1'), you would miss half of the piping needed to do the job!

The take-off sequence recommended in Chapter Twelve is used for the sample estimate:
1. Site utilities.
2. Sanitary drainage, waste and vent system.
3. Hot and cold water system.
4. Plumbing fixtures.

Before starting the actual take-off, fill in the spaces provided on the top of the quantity cost sheets. This helps avoid omissions and ensures that every sheet is accounted for.

Drainage, Waste and Vent System
Combine the building sewer (site piping), building drain, above grade waste and vent piping on one quantity sheet if you're bidding a small job. For larger, more complex jobs, use a separate sheet for

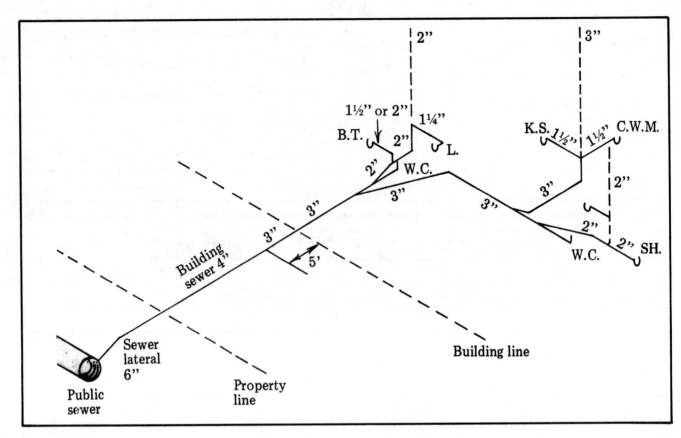

Plumbing Drainage System Illustrated and Sized on the "Flat"
Figure 14-1

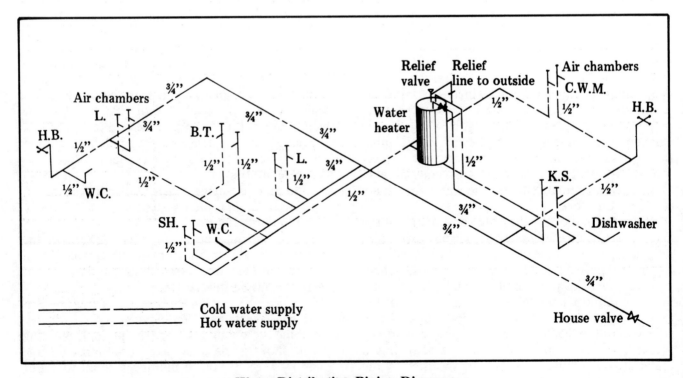

Water Distributing Piping Diagram
Figure 14-2

Taking-off & Pricing The Estimate

Center Plumbing Corp.

418 S. W. 16th Street

Hollywood, Florida 33021

(350) 579-1546

Job Address _602 N.E. 89 Street_

Sheet No. _1 of 14_ Date _2/3/84_

Estimator _J.D._ Checked By _S.A._

System Type _DWV_

Material Type _No-hub Cast Iron_

PIPING Quantity Cost Sheet

Description	Number	Unit Price	Cost
4 X 10' CAST IRON PIPE	48 LF	2.21 LF	106.08
3 X 10' CAST IRON PIPE	112 LF	1.53 LF	171.36
2 X 10' CAST IRON PIPE	34 LF	1.27 LF	43.18
		List Price	320.62
		Discount 10+10	60.92
		Net Cost	259.70

Piping Quantity Cost Sheet for DWV System
Figure 14-3

Center Plumbing Corp.

418 S. W. 16th Street

Hollywood, Florida 33021

(350) 579-1546

Job Address 602 N.E. 89 Street
Sheet No. 2 of 14   Date 2/3/84
Estimator J.D.   Checked By S.A.
System Type DWV
Material Type No-hub Cast Iron

Fitting **Quantity Cost Sheet**

Description	Number	Unit Price	Cost
4" test tee	1	7.59	7.59
4" two-way c.o. tee	1	14.07	14.07
4x3 reducer	1	3.77	3.77
3" Y	1	3.13	3.13
3" combination	1	3.77	3.77
3" 1/8 bend	1	2.01	2.01
3x2x1½x1½ angle tap tee	1	7.00	7.00
3" short sweep	1	3.10	3.10
3x2 Y	2	2.72	5.44
2" 1/8 bend	2	1.40	2.80
2" combination	2	2.40	4.80
2" short sweep	1	2.30	2.30
3x4x12 1/4 bend	2	12.30	24.60
2x1½ tapped P-trap	1	2.56	2.56
2" P-trap	1	2.26	2.26
2x1¼" tapped tee	2	2.62	5.24
4" C.O.	2	2.76	5.52
4" 1/8 bend	2	2.68	5.36
4" no-hub couplings	12	1.70	20.40
3" no-hub couplings	31	1.45	44.95
2" no-hub couplings	26	1.25	32.50
6x4" sewer sealer ring	1	13.83	13.83
		List Price	217.00
		Discount 10+10	41.23
		Net Cost	175.77

Fittings Quantity Cost Sheet for DWV System
Figure 14-4

Taking-off & Pricing The Estimate

site piping (site utilities), below grade piping, above grade waste and vent piping, and classes of piping (storm drainage, for example).

In the sample estimate the cast iron piping sizes are 4 inches, 3 inches and 2 inches. Start from some set point on the plans (generally with the largest pipe size, the building sewer in this case). With your scale, measure the length of each pipe run. Remember to tally the length of each size and type of pipe separately. Measure lengths to the extreme outside end of each pipe run—including the fitting. Not deducting the length of fittings allows for some waste. As the pipe sections are taken off, check them with a colored pencil. List each pipe size and the linear feet required on the Piping Quantity Sheet. Figure 14-3 shows the Piping Quantity and Cost Sheet.

Fittings are taken off the isometric (Figure 14-1) by checking each fitting location with a colored pencil. Fitting types, sizes and quantities are entered on the Fitting Quantity Sheet (Figure 14-4). For example: one 3-inch combination, two 2-inch 1/8 bends, one 3-inch sweep, etc.

The fixture branches (drains in some codes) are DWV copper piping. Fitting sizes are 1½ inch and 1¼ inch on this job. DWV copper piping and fittings should be listed and priced separately on the Quantity Cost Sheet. (See Figure 14-5.) Keep cast iron and copper materials separate. Each has its own discount rate in the trade price sheets you will use to compile costs.

Complete the take-off of pipe and fittings to find the total number and size of joints. Count the joints to find the number and size of no-hub couplings. Enter this number on the Fitting Quantity and Cost Sheet according to size. (See Figure 14-4.)

Labor can now be calculated from the appropriate man-hour tables in Chapter Sixteen. Labor for excavation and backfilling appears in Tables 16-35 through 16-37. The width of the trench is estimated to be 2 feet; the average depth will be 3 feet.

Labor for laying building sewer piping in open trenches is in Table 16-4.

Labor for the interior drainage piping depends on the number and size joints. Use Table 16-14 for interior drain pipe.

Labor for the fixture branch installation is in Table 16-16.

Other Drainage Systems

If the system is cast iron with lead and oakum joints, you only have to count the number and size joints. The amount of lead, oakum and gas for an installation can be calculated from the table below. Enter your estimate on the Fitting Quantity Sheet, Figure 14-4.

Many pounds of lead, oakum and gas are wasted in the installation of plumbing systems each year. This table assumes normal waste that is typical of most small jobs.

Pipe & Fitting Size in Inches	Lead In Pounds	Oakum In Pounds	Gas Tank Refills
2	1.50	.10	One tank
3	1.75	.12	for each
4	2.50	.15	2 days of
6	3.25	.18	continuous
8	6.25	.40	use.

Lead, Oakum and Gas Required to Caulk Cast-Iron Joints

If the system has cast iron pipe and fittings with neoprene joints, count the number and size of joints. Neoprene gaskets should be entered on the Fitting Quantity Sheet and priced by size. Chapter 16 has man-hour production figures for neoprene joints.

Hot and Cold Water Systems

Several types of materials are approved for water piping. The most common are galvanized pipe and copper tubing. In the sample estimate copper tubing type "L" is used.

The estimating procedure for water piping is the same as for drain, vent and waste. In small jobs combine the building water service (site utilities), the building main, and branches on one Piping Quantity Cost Sheet. For larger, more complex jobs, site piping (site utilities), building mains, risers, branches and different classes of piping (fire protection system, for example) would each be listed on separate sheets.

Take off water piping by size, type and length. Be certain that you use the same scale as was used by the architect.

In the sample estimate, the copper piping size is 3/4 inch, 1/2 inch and 3/8 inch. Start from a set point on the plans, generally with the largest pipe size. In the sample, we begin at the water meter. Scale off the lengths of pipe. As the pipe sections are listed on your take-off sheet, mark them on the plan with a colored pencil. Linear feet of pipe are listed by size on the Piping Quantity Cost Sheet. (See Figure 14-6.)

Center Plumbing Corp.

418 S.W. 16th Street

Hollywood, Florida 33021

(350) 579-1546

Job Address 602 N.E. 89 Street
Sheet No. 3 of 14 Date 2/3/84
Estimator J.D. Checked By S.A.
System Type DWV
Material Type COPPER (DWV)

Piping and Fittings Quantity Cost Sheet

Description	Number	Unit Price	Cost
1½" DWV COPPER PIPE	7 LF	1.48 LF	10.36
1¼" DWV COPPER PIPE	8 LF	1.16 LF	9.28
1½" TRAP BUSHINGS	3	.86	2.58
1¼" TRAP BUSHINGS	2	.79	1.58
1½" 90° ELLS	2	.97	1.94
1½" 45° ELLS	2	.73	1.46
1¼" 90° ELLS	2	.97	1.94
1½" TEST CAPS	3	.16	.48
1¼" TEST CAPS	2	.14	.28

List Price		29.90
Discount	25%	7.48
Net Cost		22.42

Piping and Fittings Quantity Cost Sheet for DWV System
Figure 14-5

Taking-off & Pricing The Estimate

Center Plumbing Corp.	Job Address 602 N.E. 89 Street
418 S. W. 16th Street	Sheet No. 4 OF 14 Date 2/3/84
Hollywood, Florida 33021	Estimator J.D. Checked By S.A.
(350) 579-1546	System Type WATER PIPING
	Material Type COPPER (PRESSURE)

PIPING   **Quantity Cost Sheet**

Description	Number	Unit Price	Cost
3/4" L HARD COPPER	140 LF	.82 LF	114.80
1/2" L HARD COPPER	160 LF	.61 LF	97.60
3/8" L #1 SOFT COPPER	15 LF	.42 LF	6.30
		List Price	218.70
		Discount 25%	54.68
		Net Cost	164.02

Piping Quantity Cost Sheet for Hot and Cold Water System
Figure 14-6

Fittings are taken off the piping diagram next. (See Figure 14-2.) Check each fitting with a colored pencil. Fittings should be entered on the Fittings Quantity Cost Sheet, Figure 14-7, as follows: one ¾ x ½ reducing coupling, two ¾ 45 degree ells, etc. As you count the valves and check them off with a colored pencil, enter them on the Valves Quantity Cost Sheet, Figure 14-8. For example: two ¾-inch gate valves, one ⅜-inch globe valve, etc.

As a rule of thumb, 25% of the piping cost on 3/4-inch and smaller pipe is a good estimate of the cost of fittings. This is a safe and fast method for making the fitting take-off. Fittings sized 1 inch and larger should be counted on the piping diagram, entered on the Fittings Quantity Cost Sheet and priced separately.

Miscellaneous items should be entered on the Miscellaneous Quantity Cost Sheet. (See Figure 14-9.)

The amount of solder, flux and gas for the installation can be calculated from the following table.

Pipe and Fittings Diameter Inches	Pounds in Solder	Flux in Ounces	Number of Presto Gas Tanks
⅜	1.1		
½	1.5		
¾	1.8		
1	2.3	One oz. for each pound of solder	One tank for each 3 days of continuous soldering
1¼	3.4		
1½	3.9		
2	4.6		
2½	5.4		
3	6.9		
3½	7.5		
4	8.5		

Approximate Pounds of Wire Solder, Flux and Presto Gas Tanks Estimated to Solder 100 Copper Joints

Considerable quantities of solder, flux and gas are wasted in the installation of water piping systems. This table makes allowance for the normal waste typical on most jobs.

Labor. The estimate for labor on this water piping job (Figure 14-2) is made from the following man-hour tables:

Labor for excavation and backfilling is calculated from Tables 16-35 through 16-37. Width of the trench is estimated to be 1 foot; average depth is 2 feet.

Labor for the building water service piping in an open trench installation is based on Table 16-19.

Labor for interior water piping depends on the number of water outlets (hot and cold). Table 16-23 was used.

If the water piping system is galvanized or plastic, use the same procedure as for the copper system. Man-hour tables for galvanized and plastic pipe are in Chapter 16.

Plumbing Fixtures and Equipment

Carefully review the architectural floor plan and specifications. The sample floor plans in Chapter Two show the type and quantity of plumbing fixtures. Mark each fixture with a colored pencil as you list it on your take-off sheet. This makes omissions unlikely. Enter the number, type and description of each fixture on the Fixture Quantity Sheet. (See Figure 14-10.)

Larger buildings may include equipment such as elevator sumps, sewage ejectors, or domestic pressure pumps. Each should be checked with a colored pencil as it is entered on the Equipment Quantity Sheet. Figure 14-11 is an example only; it is *not* a part of the sample estimate.

Labor for fixture installations is based on the residential plumbing fixtures table in Chapter 16.

Material Discounts

Roughing-in materials are generally priced from the latest trade price sheets or quotes provided by your plumbing wholesalers. Your dealer will probably offer you a discount which depends on the type and quantity of materials ordered. Most plumbing wholesalers offer each contractor a set discount from the manufacturer's list price. For example, cast iron and galvanized pipe and fittings may have discounts of 10% for small orders to 30% for truckload orders. Other roughing-in materials have similar discounts. Plumbing fixtures and equipment are generally quoted by the supplier without additional discounts.

Subtract all discounts from the list prices at the bottom of the Material Quantity Cost Sheets. Use the net or actual cost in summarizing the cost of this job. Chapter Fifteen shows how costs are summarized and the final bid computed.

The checklist at the end of this chapter may help you recall labor, material or other costs that might otherwise be overlooked. Remember, the first rule for every estimator is to identify every cost item and assign some dollar figure to it. Missing the actual cost of any single item by 10 or 20 percent isn't going to bankrupt anyone. But leaving something important out of the estimate can turn a nice profitable job into a major loss.

Taking-off & Pricing The Estimate

Center Plumbing Corp.
418 S. W. 16th Street
Hollywood, Florida 33021
(350) 579-1546

Job Address 602 N.E. 89 Street
Sheet No. 5 of 14   Date 2/3/84
Estimator J.D.   Checked By S.A.
System Type Hot & Cold Water
Material Type Copper (Pressure)

Quantity Cost Sheet

FITTINGS

Description	Number	Unit Price	Cost
3/4 x 1/2 Reducing Coupling	1	.31	.31
3/4" 45° Ells	2	.44	.88
3/4 x 1/2 Red. Tees	3	.47	1.41
3/4 x 3/4 x 1/2 Red. Tee	1	.47	.47
3/4 x 1/2 x 1/2 Red. Tee	1	.47	.47
3/4 x 1/2 x 1/2 Fip Tee	1	1.35	1.35
3/4 Fip x 3/4 C 90° Ell	1	1.05	1.05
3/4" 90° Ell	4	.24	.96
1/2" 90° Ell	27	.09	2.43
1/2 Fip x 1/2 C 90° Ell	1	.76	.76
3/8" 90° Ell	5	.45	2.25
3/4" Tees	4	.40	1.60
1/2" Tees	22	.16	3.52
1/2" x 3/8 Tee	1	.68	.68
3/4" Hard Caps	12	.15	1.80
1/2" Test Caps	14	.07	.98
1/2 Fip x 1/2 C Adapter	2	.45	.90
1/2 Fip x 3/8" Compression Adapter	1	.35	.35
3/8 Mip x 3/8" Compression Adapter	1	.28	.28
		List Price	22.45
		Discount 25%	5.61
		Net Cost	16.84

Fittings Quantity Cost Sheet for Hot and Cold Water System
Figure 14-7

Center Plumbing Corp.

418 S. W. 16th Street

Hollywood, Florida 33021

(350) 579-1546

Job Address *602 N.E. 89 Street*

Sheet No. *6 of 14* Date *2/3/84*

Estimator *JD* Checked By *SA*

System Type *Hot & Cold Water*

Material Type _____

Quantity Cost Sheet

VALVE

Description	Number	Unit Price	Cost
3/4" Gate Valves	2	4.56	9.12
3/8" Globe Valves	1	5.76	5.76
3/4" Hose Bibbs	2	2.69	5.38
1/2" T&P Relief Valve	1	6.87	6.87
1/2" Boiler Drains	2	1.65	3.30

List Price		30.43
Discount	30%	9.13
Net Cost		21.30

Valve Quantity Cost Sheet for Hot and Cold Water System
Figure 14-8

Taking-off & Pricing The Estimate

Center Plumbing Corp.

418 S. W. 16th Street

Hollywood, Florida 33021

(350) 579-1546

Job Address *602 N.E. 89 Street*

Sheet No. *7 of 14*   Date *2/3/84*

Estimator *J.D.*   Checked By *S.A.*

System Type _____

Material Type *MISCELLANEOUS*

Quantity Cost Sheet

MISCELLANEOUS

Description	Number	Unit Price	Cost
3" Roof Flashing	1	2.55	2.55
2" Roof Flashing	2	2.35	4.70
Solder	1 Lb.	9.06	9.06
Solder Flux	2 oz.	.88	.88
Assorted Pipe Straps	50	.11	5.50
3/4" Sleeve Material	5 LF	.09 LF	.45
1" Sleeve Material	4 LF	.13 LF	.52
Stud Guards	14	.25	3.50
Putty	5 lbs.	3.96	3.96
Pipe Compound	1 Lb.	1.65	1.65
Sand Cloth	10 FT.	.71 per	7.10
Tub Protective Cover	1	11.42	11.42
		List Price	51.29
		Discount 20%	10.26
		Net Cost	41.03

Miscellaneous Quantity Cost Sheet
Figure 14-9

Center Plumbing Corp.

418 S. W. 16th Street

Hollywood, Florida 33021

(350) 579-1546

Job Address *602 N.E 89 Street*

Sheet No. *8 of 14*   Date *2/3/84*

Estimator *J.D.*

Checked By *S.A.*

Quantity Cost Sheet — FIXTURE

Description	Number	Unit Price	Cost
TWO-PIECE WATER CLOSET W/SEATS, FLANGE PACKAGE AND SUPPLIES	2	98.49	196.98
18" OVAL LAVATORIES W/CENTERSETS, TRAPS AND SUPPLIES	2	89.46	178.92
5' RECESS CAST-IRON TUB W/DIVERTER VALVE AND WASTE	1	348.88	348.88
SHOWER COMPLETE W/VALVE, STRAINER AND 4'X6' PAN	1	67.91	67.91
24X21 STAINLESS STEEL SINK W/FAUCETS, STRAINER TRAP AND SUPPLIES	1	92.40	92.40
42 GAL. D.E. GLASS LINED REG. WATER HEATER	1	148.89	148.89
DISHWASHER UNDERCOUNTER STANDARD	1	382.00	382.00
1/2 H.P. STANDARD GARBAGE DISPOSER	1	87.00	87.00
HOT, COLD WATER AND WASTE ONLY FOR CLOTHES WASHER	1	36.57	36.57
		Net Cost	1,539.55

Figure 14-10

Taking-off & Pricing The Estimate

Center Plumbing Corp.	Job Address 602 N.E. 89 Street
418 S. W. 16th Street	Sheet No. 9 of 14 Date 2/3/84
Hollywood, Florida 33021	Estimator JD
(350) 579-1546	Checked By S.A.

EQUIPMENT — Quantity Cost Sheet

Description	Number	Unit Price	Cost
1/3 H.P. Sump Pump w/Fittings	1	192.50	192.50
20 Gallon, 40 lbs. Grease Interceptor	1	623.75	623.75
		Net Cost	816.25

Not Applicable to Sample Estimate — For Illustration Only

Figure 14-11

Center Plumbing Corp.	Job Address 602 N.E. 89 Street
418 S. W. 16th Street	Sheet No. 10 of 14 Date 2/3/84
Hollywood, Florida 33021	Estimator J.D. Checked By S.A.
(350) 579-1546	System Type

Labor Cost Summary Sheet (SANITARY SYSTEM)

Description	Number	M.H.	Total M.H.	Rate	Cost
4×10' CAST IRON PIPE	48 LF	0.41 LF	19.68	20.00	393.60
3" JOINTS (CAST IRON)	31	0.34 EA.	10.54	20.00	210.80
2" JOINTS (CAST IRON)	26	0.30 EA.	7.80	20.00	156.00
1½" JOINTS (COPPER)	14	0.32 EA.	4.48	20.00	89.60
1¼" JOINTS (COPPER)	8	0.27 EA.	2.16	20.00	43.20
SANITARY TRENCHING 2'×3'	110 LF	36.00	39.60	16.00	633.60
BACK FILLING 2'×3'	110 LF	17.00	18.70	16.00	299.20
				Total Man-hour Cost	1826.00

Figure 14-12

Center Plumbing Corp.	Job Address 602 N.E. 89 Street
418 S. W. 16th Street	Sheet No. 11 of 14 Date 2/3/84
Hollywood, Florida 33021	Estimator J.D. Checked By J.A.
(350) 579-1546	System Type

Labor Cost Summary Sheet (WATER PIPING SYSTEM)

Description	Number	M.H.	Total M.H.	Rate	Cost
3/4" L HARD COPPER	48 LF	0.21	10.08	20.00	201.60
3/4" WATER OUTLETS	1	1.25	1.25	20.00	25.00
1/2" WATER OUTLETS	18	1.23	22.14	20.00	442.80
WATER PIPING TRENCHING 1'X1'	120 LF	6.00	7.20	16.00	115.20
BACK FILLING TRENCH 1'X1'	120 LF	3.00	3.60	16.00	57.60
				Total Man-hour Cost	842.20

Figure 14-13

Center Plumbing Corp.

418 S. W. 16th Street

Hollywood, Florida 33021

(350) 579-1546

Job Address _602 N.E. 89 Street_

Sheet No. _12 of 14_ Date _2/3/84_

Estimator _J. D._ Checked By _S. A._

System Type _____

Labor Cost Summary Sheet (PLUMBING FIXTURES)

Description	Number	M.H.	Total M.H.	Rate	Cost
WATER CLOSETS COMPLETE	2	2.30	4.60	20.00	92.00
LAVATORIES COMPLETE	2	2.34	4.68	20.00	93.60
BATHTUB COMPLETE	1	4.20	4.20	20.00	84.00
SHOWER COMPLETE	1	3.50	3.50	20.00	70.00
KITCHEN SINK COMPLETE	1	2.30	2.30	20.00	46.00
DISHWASHER INSTALLATION	1	4.30	4.30	20.00	86.00
GARBAGE DISPOSAL INSTAL.	1	2.10	2.10	20.00	42.00
WATER HEATER INSTAL.	1	2.50	2.50	20.00	50.00
CLOTHES WASHER ROUGH-IN	1	1.30	1.30	20.00	26.00
				Total Man-hour Cost	573.60

Figure 14-14

Taking-off & Pricing The Estimate

Center Plumbing Corp.	Job Address 602 N.E. 89 Street
418 S. W. 16 Street	Sheet No. 14 of 14 Date 2/3/84
Hollywood, FL 33021	Estimator JD
(350) 579-1546	Checked by SA

Supplementary Costs

Based on direct job cost (including sales tax) - amount: 5572.06

Description	Rate of Discount	Cost
Tools	2%	111.44
Contingency	3%	167.16
Permits	— —	41.50
Office Overhead	6%	334.32
Sub Total Cost		6226.48
Profit	10%	622.65
Total job cost, including profit		6849.13

Figure 14-15

General Summary Check Sheet

Summary of Work
- ☐ Work under this contract
- ☐ Subcontractor's work
- ☐ Work by General Contractor
- ☐ Work by Owner
- ☐ Permits
- ☐ Sales Taxes
- ☐ Social Security Tax
- ☐ Federal Unemployment Tax
- ☐ State Unemployment Tax
- ☐ Workmen's Compensation Insurance
- ☐ Liability Insurance
- ☐ Bonding
- ☐ Superintendent
- ☐ Foreman (or Foremen)
- ☐ Sleeving and Patching

Temporary Facilities
- ☐ Water
- ☐ Sanitary Facilities
- ☐ Scaffolding
- ☐ Barricade Rental
- ☐ Construction Shed
- ☐ Construction Trailers
- ☐ Tool Rentals

Final Inspection
- ☐ Cleaning Fixtures
- ☐ Punch List
- ☐ Testing and Adjusting Plumbing Fixtures
- ☐ Testing and Adjusting Equipment

Summary of Site Work
- ☐ Trenching and Backfilling
- ☐ Building Water Service
- ☐ Water Distribution System
- ☐ Water Meter
- ☐ Building Sewer
- ☐ Sewage Collection System
- ☐ Sanitary Manholes
- ☐ Building Storm Drains
- ☐ Fire Main and Hydrants
- ☐ Cutting and Patching Sidewalks
- ☐ Cutting and Patching Paved Streets
- ☐ Irrigation Systems
- ☐ Soakage Pits and Catch Basins
- ☐ French Drains and Catch Basins
- ☐ Cesspool (where permitted)
- ☐ Grease Interceptor
- ☐ Oil Interceptor
- ☐ Lint Interceptor
- ☐ Sewage Lift Station
- ☐ Sump and Clear Water Waste Ejector
- ☐ Sump and Sewage Ejector
- ☐ Dry Well
- ☐ Building (house) Trap if Required
- ☐ Domestic Supply Well
- ☐ Irrigation Well
- ☐ Drainage Well
- ☐ Closed Well System
- ☐ Discharge Well
- ☐ Settling Tank
- ☐ Oil Retention Tank

Summary of Accessories
- ☐ Concrete Inserts
- ☐ Pipe Support Hangers
- ☐ Pipe Friction Floor Clamps
- ☐ Sleeving Material
- ☐ Pipe Insulation
- ☐ Boiler Insulation
- ☐ Hot Water Tank Insulation
- ☐ Insulating, Circulation Piping of Solar System
- ☐ Wood Backing for Shower Rods
- ☐ Wood Backing for Shower Pans
- ☐ Wood Backing for Shower Heads
- ☐ Tub Enclosures
- ☐ Shower Enclosures
- ☐ Shower Rod Straight
- ☐ Shower Rod Corner
- ☐ Above Floor Fixture Carrier
- ☐ Flue Pipe for Gas Water Heater
- ☐ Painting Metallic Pipe
- ☐ Flue Pipe for Gas Space Heater
- ☐ Hot Water Heater Relief Line
- ☐ Air Conditioning Condensate Line
- ☐ Cold Water Pipe (for electrician ground)
- ☐ Test Tee Access Covers
- ☐ Cleanout Access Covers
- ☐ Access Panels
- ☐ Valve Boxes
- ☐ Bathtub Protectors
- ☐ Grouting Fixtures
- ☐ Roof Flashings
- ☐ Pitch Pans
- ☐ Tubes Caulking Compound
- ☐ Pounds of Putty
- ☐ Pounds of Caulking Lead
- ☐ Pounds of Sheet Lead
- ☐ Pounds of Oakum or Hemp
- ☐ Presto Tank (gas refills)
- ☐ "B" Tank (gas refills)
- ☐ Compression Joint Gaskets
- ☐ No-Hub Joint Retaining Clamps
- ☐ Plastic Pipe Cement
- ☐ Plastic Pipe Cleaner
- ☐ Emery Cloth
- ☐ Soldering Paste
- ☐ Pounds Wire Solder
- ☐ Pounds of Stick Solder
- ☐ Pounds of Wiping Solder
- ☐ Copper Fitting Cleaning Brushes
- ☐ Pipe Straps
- ☐ Hanger Iron
- ☐ Stud Guards
- ☐ Pipe Thread Compound
- ☐ Cutting Oil
- ☐ Cutter Wheel Replacements
- ☐ Portland Cement
- ☐ Sand
- ☐ Gravel
- ☐ Water for Trash or Linen Chutes

Taking-off & Pricing The Estimate

General Summary Check Sheet (continued)

Summary of Interior Materials and Work
- ☐ Drainage Systems
- ☐ Drainage, Waste and Vent Piping
- ☐ Indirect Waste Piping
- ☐ Industrial Waste Piping
- ☐ Storm Water Piping
- ☐ Greasy Waste Piping
- ☐ Special Waste Piping
- ☐ Backwater Valves
- ☐ Gate Valves
- ☐ Check Valves
- ☐ Floor Drains
- ☐ Automatically Re-sealing Floor Drains
- ☐ Roof Drains
- ☐ Area Drains
- ☐ Deck Drains
- ☐ Planter Drains
- ☐ Floor Sinks
- ☐ Dilution or Neutralizing Tank
- ☐ Orthopedic Sinks, Interceptor Traps
- ☐ Cleanout Tees
- ☐ Test Tees
- ☐ Cleanouts

Domestic Water Systems
- ☐ Cold Water Piping
- ☐ Hot Water Piping
- ☐ Hot Water Return
- ☐ 180 Degree Hot Water Piping
- ☐ Blinding Valves
- ☐ Gate Valves
- ☐ Globe Valves
- ☐ Check Valves
- ☐ Angle Valves
- ☐ Hose Bibbs
- ☐ Water Softener
- ☐ Backflow Preventers
- ☐ Temperature and Pressure Relief Valves
- ☐ Control Valves
- ☐ Siphon Breakers
- ☐ Circulating Pump
- ☐ Pressure Booster Pumps

Fire Protection Systems
- ☐ Standpipe System Piping
- ☐ O.S. and Y Gate Valve
- ☐ Check Valve
- ☐ Low Pressure Cut-off Valve
- ☐ Siamese Connection
- ☐ Roof Manifold
- ☐ Standpipe Pressure Pump
- ☐ Jockey Pump
- ☐ Stairway Fire Dept. Valves
- ☐ Fire Hose Cabinets
- ☐ Fire Extinguishers
- ☐ Automatic Alarm Valve
- ☐ Sprinkler Heads

Special Piping Systems
- ☐ Chilled Water Piping
- ☐ Compressed Air Piping
- ☐ Oxygen Piping
- ☐ Vacuum Piping
- ☐ Gas Piping
- ☐ Process Piping
- ☐ Chemical Piping
- ☐ Liquid Soap Piping

Swimming Pool Systems
- ☐ Pool Piping
- ☐ Deck Drains
- ☐ Main Drain
- ☐ Scum Gutter Drains
- ☐ Filtration Equipment
- ☐ Re-circulation Inlets
- ☐ Vacuum Fitting
- ☐ Heater
- ☐ Chemical Feeder
- ☐ Control Valves
- ☐ Sight Glass
- ☐ Pressure Gauges

Plumbing Fixtures
- ☐ Water Closets
- ☐ Bidets
- ☐ Lavatories
- ☐ Bathtubs
- ☐ Showers, stall type
- ☐ Showers, pan type, tiled
- ☐ Showers, gang type
- ☐ Kitchen Sinks
- ☐ Bar Sink
- ☐ Dishwasher Undercounter
- ☐ Dishwasher Portable
- ☐ Garbage Disposer (domestic/commercial)
- ☐ Ice Maker Water Connection
- ☐ Laundry Sink
- ☐ Clothes Washer Machine, Waste, Hot and Cold Water Outlets
- ☐ Hot Water Heater
- ☐ Round Wash Sink
- ☐ Semicircle Wash Sink
- ☐ Straight Wash Sink
- ☐ Prison Lavatory
- ☐ Surgeon's Lavatory
- ☐ Surgeon's Scrub-up Sink
- ☐ Clinic Service Sink
- ☐ Eyewash Sink
- ☐ Plaster Work Sink
- ☐ Service Sink
- ☐ Urinal
- ☐ Infant's Bath
- ☐ Institutional Bath
- ☐ Emergency Bath
- ☐ Perineal Bath
- ☐ Drinking Fountain
- ☐ Grease Interceptor, inside floor mounted
- ☐ Water Softener
- ☐ Garbage Can Wash
- ☐ Pot Sink
- ☐ Glass Sink
- ☐ Scullery Sink

Center Plumbing Corp.

418 S.W. 16th Street

Hollywood, Florida 33021

(350) 579-1546

Job Address 602 N.E. 89 Street

Sheet No. 13 of 14 Date 2/3/84

Estimator J.D.

Checked By S.R.

Quantity Cost Sheet

FIXTURE

Description	Number	Unit Price	Cost
WATER CLOSETS, FLUSH VALVES, SEATS, FLANGE PACKAGES	5	102.00	510.00
18" DUAL LAVATORIES W/FAUCETS, TRAPS AND SUPPLIES	7	92.10	644.70
WALL MOUNTED URINAL, FLUSH VALVE	1	262.00	262.00
SERVICE SINK, FAUCET, 3" TRAP STANDARD	1	210.40	210.40
WALL HUNG DRINKING FOUNTAIN, TRAP, SUPPLY	1	346.50	346.50
3" FLOOR DRAINS, BRASS TOPS	2	36.20	72.40
		Net Cost	2046.00

Fixture Quantity Cost Sheet for Plumbing Core Plan on following page.
Figure 14-16

Taking-off & Pricing The Estimate

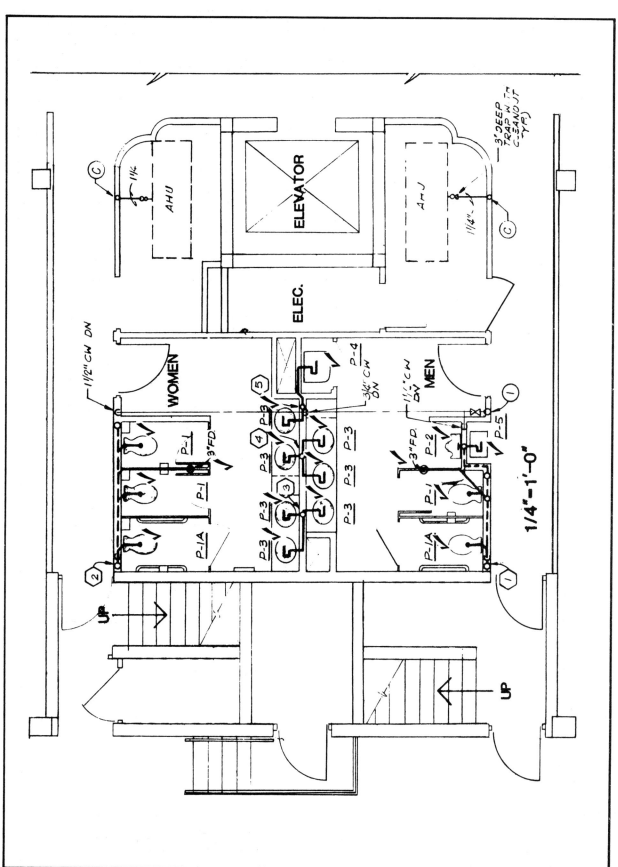

Typical Plumbing Core Plan used for preceeding Cost Sheet. Checking off each fixture helps ensure the accuracy of your estimate.
Figure 14-17

15

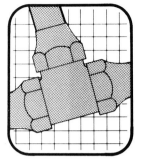

Completing The Estimate

Once you have compiled the direct cost of all labor and material, total the cost sheet for each class of work and carry the net amounts forward to the Project Summary Sheet. See Figure 15-1. Then total the items on this sheet to arrive at the *total direct job cost*. Finally add all *supplementary costs* (as outlined in this chapter) to get the *bid*, or *selling price*.

Supplementary Costs

All job and office overhead can be included in your estimate as supplementary costs. Refer to the check list in Chapter Fourteen to be sure every cost item not already included as a direct cost appears on your Project Summary Sheet. (See Figure 13-5.)

A breakdown of *supplementary costs* is shown below:

Contingency. The percentage to add as a contingency allowance varies widely. Contingencies include such items as "blue" Mondays, coffee breaks, labor problems, bad weather and remote job locations. The allowance may vary from 1 to 5% of the total direct cost, depending on the plumbing contractor's category of work and number of employees. Increase the contingency allowance where there is substantial uncertainty that your estimate will cover all costs.

Tools and miscellaneous items. One or 2% of the total direct job cost is usually considered adequate for replacement of broken, worn out and lost tools and miscellaneous supplies.

Temporary water. For simple jobs this includes only a short piece of pipe and one or two hose bibbs for use during the course of construction.

Sales tax. For new work, sales tax is generally charged on material purchases only. It should be noted that some states require you to collect sales tax on labor charges in repair and remodeling work.

Plumbing permits. Usually a set charge is assessed for each plumbing fixture, water service and sewer connection. Permit fees vary from city to city. Be sure you know what the fee is where you plan to do the work.

Job overhead. General job expenses have already been listed in the sample estimate. No additional job overhead costs are estimated in the small sample job. For larger jobs, supervision, liability insurance and other job-related expenses must be considered.

Office overhead. This is an expense on every job. Overhead expense varies widely. In the sample estimate it is assumed that office overhead expense is 6%. For many plumbing contractors the combined office and job overhead expense will be approximately 10% of the labor and material cost.

Profit. Every contractor has a different idea of a fair profit. Profit is usually computed as a percent of total labor, material and overhead cost. The goal is to end up with a reasonable return on the money invested after all bills have been paid and a reasonable salary has been received. The profit included in your bid depends on job size, competition, how badly the job is needed, and whether or not some profit is guaranteed.

Center Plumbing Corp.

418 S.W. 16th Street

Hollywood, Florida 33021

(350) 579-1546

Job Address: 602 N.E. 89 Street
Sheet No. 14 of 14 Date 2/3/84
Estimator: J.D.
Checked by: S.A.

Project Summary Sheet

Item	Classification of Work	Material	Labor	Total
1	SANITARY DRAINAGE SYSTEM	457.89	1826.00	2283.89
2	Hot and Cold Water System	202.16	842.20	1044.36
3	MISCELLANEOUS MATERIALS	41.03		41.03
4	PLUMBING FIXTURES	1539.55	573.60	2113.15
5	EQUIPMENT N/A			
	4% SALES TAX (MATERIAL ONLY)			89.63
	DIRECT JOB COST			5572.06
	TOOLS 2%			111.44
	CONTINGENCY 3%			167.16
	PERMITS			41.50
	OFFICE OVERHEAD 6%			334.32
	Subtotal Cost			6226.48
	Profit 10%			622.65
	Total Job Cost			6849.13

Figure 15-1

Cost Sheets and the Labor Cost Summary Sheets (Chapter Fourteen) are carried forward to the recapitulation sheet. (See Figure 15-1.) Total material and labor costs. Enter the combined totals in the "total" column. Each column should be added until all direct job costs have been entered and totaled. Complete the estimate by adding all supplemental costs. The sum of these figures is the *selling price*.

Writing Up The Contract

You are now ready to submit your proposal to the owner or general contractor. Be certain that the contract includes and excludes everything included and excluded from your estimate. When the contract is signed you are legally bound. Chapter Eleven gives detailed information on contracts.

16

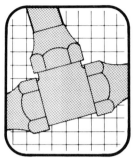

Man-hour Tables For Plumbing

A Plumber's Normal Hour
Many factors influence how much plumbing work a man can finish in an hour. These include:
- The quality of job supervision
- Job conditions
- Efficiency of the general contractor
- Quantity and quality of tools and equipment
- General economic conditions
- Weather conditions

The man-hour production tables in this chapter are based on the efficiency of an average plumber working under normal conditions. Your judgment and experience are needed to evaluate whether the job in question involves "normal" or "unusual" working conditions.

Labor Cost
No dollar costs are listed here. Labor costs vary from area to area and from year to year. But you should know the actual cost per hour for each classification of worker on your payroll. Your cost includes the basic wage, employer's contribution to welfare, pension, vacation, and apprentice funds, taxes and insurance expenses.

How To Use The Tables
The installation times in Tables 16-1 through 16-41 are based on single items of material in single job operations.

These tables give the time needed to install each component of nearly any plumbing system. The tables should reduce the amount of counting you have to do.

Chapter Fourteen explained how to use the tables based on the number, size and type of joints, the number and type of liquid piping outlets, the linear feet of pipe, the number of hangers and supports, and the number and type of plumbing fixtures, equipment and appliances.

For example, assume you have listed the number of residential tank type floor-mounted water closets on the material take-off sheet. Multiply the number of water closets by the time period in the appropriate table. Then multiply this figure by the hourly labor cost. This yields the total labor cost to install all water closets in a building.

The titles and notes on the tables are self-explanatory and should present no problem if you are familiar with plumbing practice.

The tables in this chapter should help you produce accurate labor estimates that avoid guesswork.

Building Piping Systems
Linear Feet Drainage Piping
Man-hours listed for installing piping materials and fittings *include* delivery and handling at the job site. Man-hours *do not include* excavation, backfilling or supports. Other tables in this chapter cover those subjects.

	Man-Hours Per Linear Foot		
Pipe Size	Open Trench Installation	Horizontal Installation	Vertical Installation
2"	0.34	0.38	0.42
3"	0.39	0.43	0.47
4"	0.44	0.49	0.54
6"	0.58	0.64	0.70
8"	0.76	0.85	0.94

Type: Lead and Oakum Joints Cast-iron piping with fittings in five foot lengths. Man-hours for piping installations exceeding 20 feet developed length. Usually used in building sewers, building drains, and storm water piping systems

Table 16-1

	Man-Hours Per Linear Foot		
Pipe Size	Open Trench Installation	Horizontal Installation	Vertical Installation
2"	0.31	0.35	0.40
3"	0.36	0.40	0.44
4"	0.41	0.46	0.51
6"	0.55	0.61	0.67
8"	0.74	0.82	0.90

Type: Lead and Oakum Joints Cast-iron piping with fittings in ten foot lengths. Man-hours for piping installations exceeding 20 feet developed length. Usually used in building sewers, building drains, and storm water piping systems

Table 16-2

	Man-Hours Per Linear Foot		
Pipe Size	Open Trench Installation	Horizontal Installation	Vertical Installation
2"	0.32	0.36	0.40
3"	0.37	0.41	0.45
4"	0.42	0.47	0.52
6"	0.56	0.62	0.68
8"	0.74	0.83	0.92

Type: No-Hub Joints Cast-iron piping with fittings in five foot lengths. Man-hours for piping installations exceeding 20 feet developed length. Usually used in building sewers, building drains, and storm water piping systems

Table 16-3

	Man-Hours Per Linear Foot		
Pipe Size	Open Trench Installation	Horizontal Installation	Vertical Installation
2"	0.31	0.35	0.39
3"	0.36	0.39	0.43
4"	0.41	0.45	0.49
6"	0.55	0.60	0.64
8"	0.72	0.81	0.85

Type: No-Hub Joints Cast-iron piping with fittings in ten foot lengths. Man-hours for piping installations exceeding 20 feet developed length. Usually used in building sewers, building drains, and storm water piping systems

Table 16-4

	Man-Hours Per Linear Foot		
Pipe Size	Open Trench Installation	Horizontal Installation	Vertical Installation
2"	0.30	---	---
3"	0.34	---	---
4"	0.39	---	---
6"	0.52	---	---
8"	0.72	---	---

Type: Compression Joints Cast-iron piping with fittings in five foot lengths. Man-hours for piping installations exceeding 20 feet developed length. Usually used in building sewers and storm water drains, sitework only

Table 16-5

	Man-Hours Per Linear Foot		
Pipe Size	Open Trench Installation	Horizontal Installation	Vertical Installation
2"	0.29	---	---
3"	0.33	---	---
4"	0.37	---	---
6"	0.49	---	---
8"	0.68	---	---

Type: Compression Joints Cast-iron piping with fittings in ten foot lengths. Man-hours for piping installations exceeding 20 feet developed length. Usually used in building sewers and storm water drains, sitework only

Table 16-6

Pipe Size	Man-Hours Per Linear Foot		
	Open Trench Installation	Horizontal Installation	Vertical Installation
6"	0.35	---	---
8"	0.39	---	---
10"	0.41	---	---

Type: Compression Joints Asbestos cement piping with fittings in five foot lengths. Man-hours for piping installations exceeding 20 feet developed length. Usually used in storm water and subsoil drains, sitework only.

Table 16-7

Pipe Size	Man-Hours Per Linear Foot		
	Open Trench Installation	Horizontal Installation	Vertical Installation
6"	0.34	---	---
8"	0.37	---	---
10"	0.39	---	---

Type: Compression Joints Asbestos cement piping with fittings in ten foot lengths. Man-hours for piping installations exceeding 20 feet developed length. Usually used in storm water and subsoil drains, sitework only.

Table 16-8

Pipe Size	Man-Hours Per Linear Foot		
	Open Trench Installation	Horizontal Installation	Vertical Installation
4"	0.41	---	---
6"	0.43	---	---
8"	0.48	---	---
10"	0.52	---	---

Type: Compression Joints Vitrified clay piping with fittings in two and one-half or three foot lengths. Man-hours for piping installations exceeding 20 feet developed length. Usually used in building sewers, storm water and subsoil drains, sitework only.

Table 16-9

Pipe Size	Man-Hours Per Linear Foot		
	Open Trench Installation	Horizontal Installation	Vertical Installation
3"	0.35	---	---
4"	0.39	---	---
6"	0.41	---	---

Type: Tapered Friction Joints Bituminized piping with fittings in eight foot lengths. Man-hours for piping installations exceeding 20 feet developed length. Usually used in building sewers, storm water and subsoil drains, sitework only.

Table 16-10

Pipe Size	Man-Hours Per Linear Foot		
	Open Trench Installation	Horizontal Installation	Vertical Installation
1¼"	0.09	0.11	0.12
1½"	0.12	0.14	0.15
2"	0.14	0.17	0.18
2½"	0.17	0.19	0.21
3"	0.20	0.23	0.26
4"	0.25	0.28	0.31
6"	0.31	0.34	0.38

Type: Solvent Cemented Joints Schedule 40 plastic DWV piping with fittings in twenty foot lengths. Man-hours for piping installations exceeding 20 feet developed length. Usually used in sanitary drainage, storm water drainage, chemical and acid drainage systems. May be used for site and interior work where permitted by code.

Table 16-11

Pipe Size	Man-Hours Per Linear Foot		
	Open Trench Installation	Horizontal Installation	Vertical Installation
1¼"	0.26	0.29	0.31
1½"	0.29	0.32	0.35
2"	0.34	0.38	0.42
3"	0.44	0.49	0.54
4"	0.54	0.60	0.66
5"	0.69	0.77	0.85
6"	0.84	0.94	1.03

Type: Soldered Joints Copper DWV piping with fittings in twenty foot lengths. Man-hours for piping installations exceeding 20 feet developed length. Usually used in interior drainage systems.

Table 16-12

Interior Drainage Piping

Man-hours listed are for roughing-in drainage pipe for fixtures within a 10-foot radius of the fixture. Unlike drainage tables in which production is measured in man-hours per linear foot, this table is based on man-hours per joint. Times include delivery, handling, laying-out, notching and securing piping in the partitions.

Pipe Size	Man-hours Per Joint
2"	0.32
3"	0.36
4"	0.41
6"	0.54

Type: Lead and Oakum Joints.
Cast-iron piping with fittings
Table 16-13

Pipe Size	Man-hours Per Joint
2"	0.30
3"	0.34
4"	0.39
6"	0.51

Type: No-Hub Joints. Cast-iron piping with fittings
Table 16-14

Pipe Size	Man-hours Per Joint
1¼"	0.12
1½"	0.14
2"	0.17
2½"	0.19
3"	0.23
4"	0.29
6"	0.34

Type: Solvent Cemented Joints.
Plastic DWV piping with fittings
Table 16-15

Pipe Size	Man-hours Per Joint
1¼"	0.27
1½"	0.32
2"	0.38
3"	0.49
4"	0.60
5"	0.75
6"	0.89

Type: Soldered Joints. Copper DWV
piping with fittings
Table 16-16

Piping for Liquids and Gases

The man-hours listed are for installing piping and fittings and include delivery and handling at the job site. Man-hours do not include excavation, backfilling or supports.

Pipe Size	Man-Hours Per Linear Foot		
	Open Trench Installation	Horizontal Installation	Vertical Installation
½"	0.18	0.20	0.21
¾"	0.21	0.23	0.24
1"	0.23	0.26	0.28
1¼"	0.26	0.29	0.30
1½"	0.27	0.30	0.32
2"	0.28	0.31	0.33
2½"	0.32	0.36	0.38
3"	0.35	0.39	0.41
4"	0.39	0.43	0.45
5"	0.45	0.49	0.53
6"	0.49	0.54	0.56

Type: **Black Steel, Galvanized Steel, and Wrought Iron Piping in twenty-one foot lengths** Includes threaded joints and fittings. Man-hours for piping installations exceeding 20 feet developed length. Usually used in building sitework, building mains and building risers.

Table 16-17

Pipe Size	Man-Hours Per Linear Foot		
	Open Trench Installation	Horizontal Installation	Vertical Installation
¼"	---	---	---
⅜"	---	---	---
½"	0.16	---	---
¾"	0.17	---	---
1"	0.19	---	---
1¼"	0.25	---	---

Type: Soft Copper Tubing K and L in sixty foot coil Includes soldered joints and fittings. Man-hours for piping installations exceeding 20 feet developed length. Usually used in building sitework and building main. ¼" to ⅜" special installations.

Table 16-18

Pipe Size	Man-Hours Per Linear Foot		
	Open Trench Installation	Horizontal Installation	Vertical Installation
½"	0.16	0.18	0.19
¾"	0.21	0.23	0.24
1"	0.22	0.25	0.27
1¼"	0.24	0.27	0.29
1½"	0.25	0.28	0.30
2"	0.27	0.30	0.32
2½"	0.32	0.36	0.38
3"	0.35	0.39	0.41

Type: Hard Copper tubing K, L and M in twenty foot lengths Includes soldered joints and fittings. Man-hours for piping installation exceeding 20 feet developed length. Usually used in building sitework, building main and building risers.

Table 16-19

Pipe Size	Man-Hours Per Linear Foot		
	Open Trench Installation	Horizontal Installation	Vertical Installation
¾"	0.18	---	---
1"	0.19	---	---
1¼"	0.21	---	---
1½"	0.23	---	---
2"	0.25	---	---
2½"	0.29	---	---
3"	0.32	---	---

Type: ABS, PVC and PE Plastic piping in twenty foot lengths (Rated 160 p.s.i.) Includes cemented and compression joints and fittings. Man-hours for piping installation exceeding 20 feet developed length. Usually used in building sitework only where permitted by code.

Table 16-20

Interior Pressure Piping

Man-hours are for roughing-in piping outlets within a 10-foot radius of the fixture, equipment or appliance. Estimates are based on man-hours per outlet. This includes delivery, handling, lay-out, notching and securing piping to the partition.

Pipe Size	Man-hours Per Outlet
½"	1.25
¾"	1.27
1"	1.32
1¼"	1.73
1½"	1.75
2"	1.85
2½"	2.10
3"	2.25
4"	3.15
5"	3.30
6"	3.75

Type: Threaded Joints. Black Steel, Galvanized Steel and Wrought Iron Piping with Fittings
Table 16-21

Tubing Size	Man-hours Per Outlet (4 to 6 feet, Special Installation)
1/4"	1.12
3/8"	1.14
1/2"	1.25

Type: **Soldered, Flare or Compression Joints.** Soft copper tubing type L with fittings
Table 16-22

Tubing Size	Man-hours Per Outlet
1/2"	1.23
3/4"	1.25
1"	1.29
1 1/4"	1.65
1 1/2"	1.67
2"	1.75

Type: **Soldered Joints.** Hard copper tubing type K, L and M with fittings
Table 16-23

Pipe Size	Man-hours Per Outlet
1/2"	1.20
3/4"	1.23
1"	1.27

Type: **Cemented Joints** High-temperature vinyl CPVC plastic pipe with fittings. Where permitted by code, can be used above first floor slab. For interior use only.

Table 16-24

Control Valves

Valve Size	Man-hours Per Valve
1/4"	0.26
3/8"	0.26
1/2"	0.31
3/4"	0.32
1"	0.36
1 1/4"	0.40
1 1/2"	0.47
2"	0.72
2 1/2"	0.86
3"	1.20
4"	1.41

Type: **Threaded Connection** Gate, Globe, Angle or Check Valves, Man-hours for new work only.

Table 16-25

Valve Size	Man-hours Per Valve
1/4"	0.25
3/8"	0.25
1/2"	0.30
3/4"	0.31
1"	0.34
1 1/4"	0.39
1 1/2"	0.41
2"	0.69

Type: **Soldered Joint** Gate, Globe or Check Valves. Man-hours for new work only.

Table 16-26

Valve Size	Man-hours Per Valve
2"	1.35
2 1/2"	1.65
3"	2.26
4"	3.41
5"	4.15
6"	5.06

Type: **Flanged Joint** Gate or Check Valves. Man-hours for new work only.

Table 16-27

Man-hour Tables For Plumbing

Valve Size	Man-hours Per Valve
¼"	0.27
⅜"	0.27
½"	0.33
¾"	0.35
1"	0.38
1¼"	0.43
1½"	0.44
2"	0.68

Type: Threaded Joint Lever handle or square head gas cocks. Man-hours for new work only.

Table 16-28

Valve Size	Man-hours Per Valve
½"	0.49
¾"	0.52
1"	0.55

Type: Threaded Joint Hot water heater relief valves. Man-hours for new work only.

Table 16-29

Pipe Size	Man-hours Per Breaker
½"	0.49
¾"	0.52
1"	0.58
1¼"	0.62
1½"	0.64

Type: Threaded Joint. Vacuum Breaker Man-hours for new work only.

Table 16-30

Pipe Size	Man-hours Per Absorber
¾"	0.48
1"	0.66
1¼"	0.75
1½"	0.76
2"	1.22

Type: Threaded Joint. Shock Absorber Man-hours for new work only.

Table 16-31

Pipe, Painting, Insulation and Sleeves

Pipe Size	Man-hours Per Linear Foot
½"	0.19
¾"	0.20
1"	0.21
1¼"	0.22
1½"	0.22
2"	0.23
2½"	0.24
3"	0.26
4"	0.28
5"	0.32
6"	0.35

Pipe Painting
Table 16-32

Pipe Size	Man-hours Per Linear Foot
½"	0.41
¾"	0.42
1"	0.45
1¼"	0.49
1½"	0.52
2"	0.61
2½"	0.66
3"	0.71
4"	0.72
5"	0.74
6"	0.79

Pipe Insulation
Table 16-33

Pipe Size	Man-hours For Each
½"	0.15
¾"	0.17
1"	0.23
1¼"	0.28
1½"	0.32
2"	0.36
2½"	0.44
3"	0.45
4"	0.49
5"	0.51
6"	0.53

Plastic or Steel Pipe Sleeves
Table 16-34

Hand Excavation

Man-hours are for picking and loosening light and medium soil and placing soil on the bank of the trench. Table figures are in laborer hours per 100 linear feet of pipe trench.

Width In Feet	Depth In Feet			
	1	2	3	4
1	6.00	12.00	18.00	24.00
2	12.00	24.00	36.00	48.00

Table 16-35

Use power equipment where soil consists of heavy clay, gravel or rock. Hand excavate to remove loose dirt and shape trench bottom to receive pipe. Man-hours are per 100 linear feet.

Width in Feet	Man-hours
1	2.20
2	5.10
3	12.25
4	26.70

Table 16-36

Man-hours are for backfilling trenches by hand and do not include compacting loose dirt. Figures are per 100 linear feet.

Width In Feet	Man-hours Depth in Feet			
	1	2	3	4
1	3.00	5.75	8.00	12.00
2	5.75	11.50	17.00	22.00

Table 16-37

Horizontal Supports (see Figure 16-43)

Pipe Size	Man-hours for Each
½"	0.15
¾"	0.16
1"	0.19
1¼"	0.19
1½"	0.21
2"	0.22
2½"	0.24
3"	0.26
4"	0.33
5"	0.37
6"	0.51

Type: Split Ring and Clevis Type Hangers With Rod and Inserts.
Horizontal Pipe Supports
Table 16-38

Vertical Supports (see Figure 16-44)

Pipe Size	Man-hours for Each
1¼"	0.22
1½"	0.24
2"	0.26
2½"	0.29
3"	0.33
4"	0.37
5"	0.41
6"	0.53

Type: Floor Clamp Type With Nuts and Bolts.
Vertical Pipe Supports
Table 16-39

Plumbing Fixtures

Man-hours are to install plumbing fixtures, special fixtures, fixture trim, accessories and equipment in residential and commercial buildings. This includes delivery and handling at the job site. Man-hours do not include rough-in piping or fixture carriers where required. They are for installations and connection to existing outlets.

Man-hour Tables For Plumbing

Item	Man-hours for Each
Lavatory	
Countertop	2.34
Wall-hung	2.94
With legs and towel bars	3.04
Pedestal	3.14
Water Closet	
Flush valve, floor-mounted	2.30
Flush valve, wall-mounted	3.00
Tank type, floor-mounted	2.30
Tank type, wall-mounted	3.10
Tank type, tank wall-mounted	3.15
One piece, floor-mounted	2.90
One piece, wall-mounted	3.00
Concealed carrier installation	2.10
Shower	
Stall type prefabricated, metal	4.90
Stall type one piece, fiberglass	4.50
Base preformed, one piece	3.20
Tiled, pan cut to size	3.50
Valve body only	1.10
Curtain rod, straight	1.00
Door enclosure	3.10
Bathtub	
Leg type	4.90
Recessed	4.20
Corner	5.10
Combination, one piece fiberglass	6.50
Shower curtain rod, straight	1.00
Shower curtain rod, corner	1.20
Shower door enclosure	3.15
Kitchen Sink	
Single bowl countertop	2.30
Double bowl countertop	2.70
Triple bowl countertop	3.50
Bar sink countertop	2.15
Laundry Sink	
Single bowl countertop	2.30
Double bowl countertop	2.70
Single bowl with stand	2.40
Double bowl with stand	2.80
Single bowl with pedestal	2.75
Hot Water Heater	
Regular 20-42 gallon electric	2.50
Undercounter 20-30 gallon electric	3.10
Table top 20-30 gallon electric	2.80
Regular 20-42 gallon gas (less flue)	3.50
Special Fixtures and Connections	
Bidet	3.25
Garbage disposer	2.10
Dishwasher	4.30
Water supply connected to refrigerator ice maker	2.25
Clothes washer, waste, hot and cold water outlets	1.30
Gas, hot water heater flue pipe 3" or 4", 10' maximum	3.20
Water softener	4.15
Air conditioning waste from pan to drain pipe	1.10
Hose bibbs	.80
Grease interceptor 8 lb. capacity	4.00
Grease interceptor 14 lb. capacity	4.50
Grease interceptor 20 lb. capacity	5.75
Grease interceptor 30 lb. capacity	6.10

Residential Plumbing Fixtures
Table 16-40

Item	Man-hours for Each
Lavatory	
Countertop	2.34
Wall-hung	2.34
Wheelchair	3.75
Barber and beauty parlor	3.60
Prison	3.40
Patient's	4.10
Surgeon's, remote knee operated	5.20
Surgeon's, remote pedal operated	5.20
Concealed carrier installation	2.10
Sink	
Wash-up, 36 inch round	5.50
Wash-up, 36 inch semicircle	6.25
Wash-up, 54 inch round	7.10
Wash-up, 4 foot straight	5.60
Wash-up, 6 foot straight	6.50
Surgeon's scrub-up	5.20
Clinical	3.75
Eyewash	4.80
Service, regular	3.25

Commercial Plumbing Fixtures
Table 16-41

Item	Man-hours for Each
Mop receptacle	4.10
Plaster with interceptor trap	3.00
Developing with dilution tank	3.00
Pot, single compartment	4.00
Pot, double compartment	4.15
Pot, triple compartment	4.30
Pot, scullery	3.20
Glass, rinse	3.10
Floor type	4.50
Water Closet	
Flush valve, floor-mounted	2.30
Flush valve, wall-mounted	3.00
Tank type, floor-mounted	2.30
Tank type, wall-mounted	3.10
Tank type, tank wall mounted	3.15
Outpatient, pedal operated	4.20
Concealed carrier installation	2.10
Urinal	
Wall-hung	5.50
Stall-type	4.00
Floor-mounted	3.10
Trough with flushing tank	6.12
Bath	
Restal recess	4.20
Infant	5.10
Institutional	5.30
Emergency	5.30
Perineal	6.10
Drinking Fountain	
Wall-mounted	5.30
Semi-recessed	5.10
Free standing	4.20
Bubbler	2.15
Hot Water Heater	
Regular 52-60 gallon electric	3.30
Regular 75 gallon electric	4.20
Regular 100-120 gallon electric	6.10
Regular 60-75 gallon gas (less flue)	5.20
Regular 100-120 gallon gas (less flue)	7.00
Hot Water Generator	
500 G.P.H. electric	26.00
1000 G.P.H. electric	32.00
1500 G.P.H. electric	41.00
500 G.P.H. gas	29.00
1000 G.P.H. gas	34.00
1500 G.P.H. gas	42.00
500 G.P.H. oil	29.00
1000 G.P.H. oil	34.00
1500 G.P.H. oil	42.00

Item	Man-hours for Each
Sewage Ejector Pumps	
Single type 30 G.P.M.	32.60
Duplex type 50 G.P.M.	37.50
Clear Water Sump Pumps	
Single type 30 G.P.M.	28.00
Duplex type 30 G.P.M.	32.40
Constant Pressure Pump w/Control Panel	
Domestic single type	25.10
Domestic duplex type	29.30
Constant Pressure Pump w/Control Panel	
Fire 500 G.P.M.	55.00
Fire 750 G.P.M.	59.00
Fire jockey 15 pounds per square inch	12.30
Circulating Pumps	
$1/12$ to $1/4$ horsepower	6.30
$1/3$ to $1/2$ horsepower	7.50
$3/4$ to 1 horsepower	9.30
Backflow Preventers	
Hose bibb type	.45
1 inch in line type	1.80
1 1/4 inch in line type	2.45
1 1/2 inch in line type	2.55
2 inch in line type	2.90
Grease Interceptors	
100 lb. capacity	15.10
150 lb. capacity	17.50
200 lb. capacity	20.30
400 lb. capacity	33.10
700 lb. capacity	57.20
1000 lb. capacity	69.00
1200 lb. capacity	74.50
Oil Interceptors	
50 G.P.M. flow rate	7.50
100 G.P.M. flow rate	15.30
200 G.P.M. flow rate	22.00
300 G.P.M. flow rate	29.10
500 G.P.M. flow rate	45.20
Fire Protection	
Hose cabinets (recessed 125' hose)	7.00
4 inch Siamese connection	4.75
6 inch Siamese connection	5.75
4 inch roof manifold	4.25
6 inch roof manifold	5.25
2 1/2 inch fire department valve	1.50
Acid Neutralizing Tanks	
15 gallon capacity	4.50
35 gallon capacity	7.25

Commercial Plumbing Fixtures
Table 16-41 (continued)

Item	Man-hours for Each	Item	Man-hours for Each
50 gallon capacity	8.50	4-inch can wash drain	2.25
100 gallon capacity	9.75	2-inch floor drain with integral trap	2.10
Special Interceptor Tanks		3-inch floor drain with integral trap	2.40
15 G.P.M. flow rate	4.50	4-inch floor drain with integral trap	2.65
20 G.P.M. flow rate	6.10	2-inch floor drain with sediment bucket	2.20
30 G.P.M. flow rate	7.25	3-inch floor drain with sediment bucket	2.50
50 G.P.M. flow rate	8.20	4-inch floor drain with sediment bucket	2.75
Special Fixtures and Drain Connections		2-inch roof drain	2.30
4-inch back water valve	4.65	3-inch roof drain	2.40
Garbage disposer	2.20	4-inch roof drain	2.50
Dishwasher	4.60	5-inch roof drain	2.60
Cleanout access cover	.70	6-inch roof drain	2.80
Valve access cover	.70	Roof drain sheet flashings	2.20
3-inch can wash drain	2.15	Pitch pans	1.25

Commercial Plumbing Fixtures
Table 16-41 (continued)

Minutes	Hours	Minutes	Hours	Minutes	Hours	Minutes	Hours
1	.017	16	.267	31	.517	46	.767
2	.033	17	.283	32	.533	47	.783
3	.050	18	.300	33	.550	48	.800
4	.067	19	.317	34	.567	49	.817
5	.083	20	.333	35	.583	50	.833
6	.100	21	.350	36	.600	51	.850
7	.117	22	.367	37	.617	52	.867
8	.133	23	.383	38	.633	53	.883
9	.150	24	.400	39	.650	54	.900
10	.167	25	.417	40	.667	55	.917
11	.183	26	.433	41	.683	56	.933
12	.200	27	.450	42	.700	57	.950
13	.217	28	.467	43	.717	58	.967
14	.233	29	.483	44	.733	59	.983
15	.250	30	.500	45	.750	60	1.000

Minutes to Decimal Hours Extended to Nearest Fraction
Table 16-42

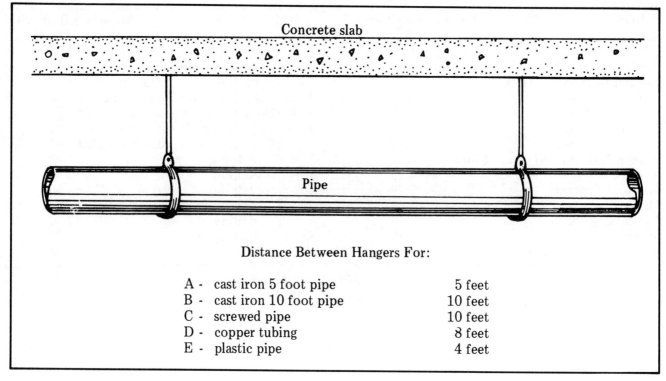

Horizontal Supports
Figure 16-43

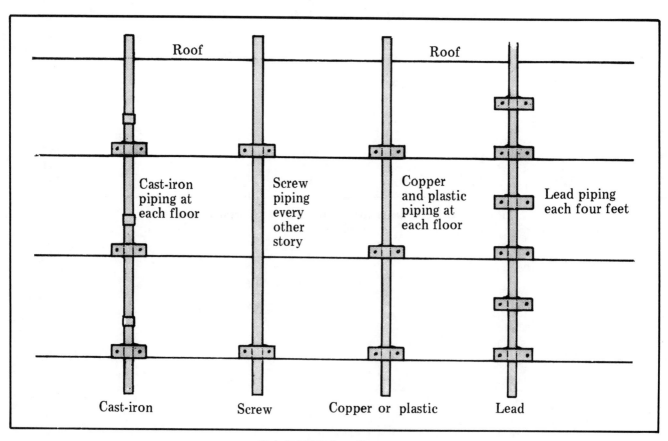

Vertical Piping Supports
Figure 16-44

Part 4

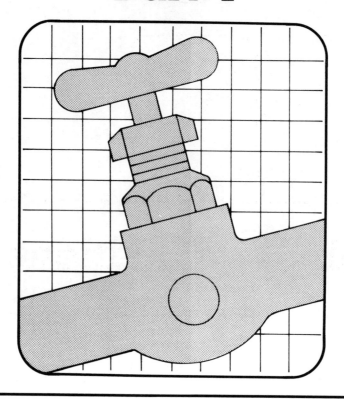

Appendix

- Job Specifications
- Calculating Water Supply & Fixture Requirements
- Making Plumbing Calculations
- Definitions
- Standard Abbreviations
- Common Fixture, Plumbing, & Fitting Symbols

A

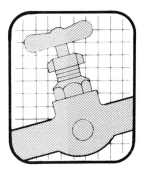

Job Specifications

The typical specifications for residential or simple commercial jobs are presented in Section I. Note that it is the owner's prerogative to select the type and color of plumbing fixtures and their trim. The architect may give certain general requirements for the type of roughing-in materials and the installation methods, but usually he will leave the specifics to the plumbing contractor.

The typical specifications for large commercial buildings are presented in Section II. Note that the architect and engineer often provide very detailed specifications regarding fixtures, trim, equipment, components, roughing-in materials, and even installation methods. The architect or engineer usually supervises the complete installation of all plumbing systems in the building. Any deviation from specifications must be in writing and approved before changes can be made.

Section I
Typical Specifications for a Simple Job
Plumbing

1. All work is to be in accordance with the local plumbing codes.

2. Building drainage system design is based on 1/8-inch minimum fall, and deviations shall be approved by the architect and engineer.

3. All underground valves are to be installed in precast concrete boxes.

4. The plumbing contractor is to furnish all required flashings for pipes to the roofing contractor.

5. Materials shall be all new as follows:
   A. Drainage piping—No hub cast iron, standard weight.
   B. Water piping—Copper, type L.
   C. A/C condensate drain—Schedule 40 PVC, plastic.
   D. Plumbing fixtures—Type, color and trim selected by owner.
   E. Electric water heaters—glass lined, 5-year guarantee.
   F. Valves—125 psi.
   G. Hose bibbs—equipped with backflow preventer.

6. The plumbing contractor shall provide all excavation and backfill compaction required for all plumbing underground work.

7. He shall perform the following tests:
   A. Water piping shall be subjected to hydrostatic pressure test of 100 psi for a period of time sufficient to examine the entire system, but not less than one hour.
   B. Drainage piping—The ends of the system shall be plugged and the system filled with a 5-foot head of water. Water shall stand until the inspection is made and the water level remains constant.

C. Correct all the defects disclosed by the above tests.

8. He shall sterilize all water lines with a mixture of 2 pounds of chlorinated lime to each 1,000 gallons of water (50 p.p.m. of available chlorine), retain the mixture in pipes 24 hours and flush it thoroughly with potable water before placing it in service.

9. Complete system—Fixtures and equipment shall be given an in-service test after completion of the installation.

10. The plumbing contractor shall furnish a written guarantee that all plumbing work shall be free from defects of materials and workmanship for a period of one year, from the date of final acceptance, and that he will, at his expense, repair and replace all work which becomes defective during the guarantee period.

Section II
Typical Specifications for a Large Commercial Building
Section 16A—Plumbing

General

1.01 Supplementary Documents

Contract Conditions and Division 1 Requirements apply to work specified in this Section.

1.02 Scope

Provide all necessary labor, materials, equipment, accessories, transportation, and services required for the complete installation of all Plumbing Systems in the building shown on the drawings. For convenience, drawings showing primarily Plumbing have been numbered with "P."

1.03 Codes

Make the installation in a manner that will comply with applicable codes and laws. Where the requirements of Contract Documents exceed Code requirements, comply with Contract Documents.

1.04 Coordination

Coordinate work under this Division with work under all other Divisions so that all components will be installed in the proper place at the proper time, avoiding conflicts between trades and improper space for servicing. Components improperly installed shall be removed and relocated as directed by the architect, at no additional cost to the owner.

1.05 Drawings

Locate all items of equipment by on-the-job measurements and coordinate with the other trades. Under no circumstances shall the drawings be scaled. Model numbers in the specifications or schedules on drawings are not intended to designate all the required trim.

1.06 General Requirements For All Equipment

A. Provide all necessary parts and accessories, even though the parts and accessories are not specifically mentioned herein.

B. Provide a factory applied finish to all exterior surfaces. Any items which have the finish marred must have touch up or be refinished to a new condition before final acceptance.

C. Furnish three copies of spare parts lists, and operating and maintenance instructions. Bind them in folders with suitable identifications on the front cover. Deliver them to the architect at the time of final acceptance.

D. Rotating parts must be in static and dynamic balance.

1.07 Noise

Eliminate any abnormal noises, which are not considered by the architect to be an inherent part of the system as designed. Rattling equipment, piping and squeaks in rotating equipment components will not be acceptable.

1.08 Submittals

Submittals shall show sufficient data (sizes, capacities, construction, methods, materials, finishes, etc.) to indicate compliance with the Contract Documents. Provide submittals on the following:

    01 Domestic water pumps, tanks and controls
    02 Controls and wiring diagrams
    03 Insulation
    04 Water heaters
    05 Fixtures and trim
    06 Drains and specialties
    07 Fire hose cabinets, valves and racks
    08 Fire protection specialties
    09 Fire pumps and controls

1.09 Substitutions

The names and model numbers of manufacturers listed in the Contract Documents have been used to establish a standard of quality. Provide these as indicated unless prior approval is obtained in writing.

1.10 Record Drawings

A. Provide a set of prints at the jobsite and periodically indicate all changes and means of access to all equipment and valves requiring service.

B. Provide a set of erasable sepias, and draw all changes noted on the working prints. Mark

Job Specifications

each sheet "As-Built" and deliver to the architect.

Section 16P—Fire Protection Systems

Part I—General

1.01 Supplementary Documents
Contract Conditions and Division 1 requirements apply to work specified in this Section.

1.02 General Requirements:
  A. Provide a complete fire protection system, in accordance with NFPA and Code requirements. All equipment, valves, and other devices shall be U.L. approved for fire services.
  B. All of the stipulations govern this Section, except as modified herein.

Part II—Materials and Methods

2.01 Standpipe System Piping
  A. Pipe: Schedule 40, black steel, screwed or welded fittings, inside building. Cast iron, class 150 A.W.W.A. with mechanical joints underground.
  B. Testing: Test all components of the system at 50 psi more than the maximum operating pressure, but in no case less than 200 psi test pressure. Remake all joints which leak, and repeat the test until no leaks are shown.
  C. Hose equipment: Models listed are Elkhart.
    01 Cabinets: Elkhart series 2634R recessed with solid steel door with raised letters "FIRE HOSE," prime coated exterior, white baked enamel interior, continuous hinge, lever handle, cam catch. Provide hose racks only for the parking garage.
    02 Trim for each Cabinet and each hose rack:
      a. No. 1.5—100 unit with 1½-inch angle valve, red enamel rack, brass nipple, 100 ft. of linen hose, lug coupling, hose to pipe adapter, and No. HN-4-L1½-inch Lexan Fog Nozzle.
      b. Provide drain/vent valve with each rack.
      c. Provide pressure gauge at top of each riser.
    03 Fire department connections: Elkhart No. 4-25, 2½-inch valve with cap and chain.
    04 Siamese building: Elkhart No. 29 Series chrome plated sidewalk unit with three independent self-closing clapper valves, plugs and chains lettered "Autospk" and "Standpipe," sizes 6-inch x 2½-inch x 2½-inch, ball drip.
    05 Siamese garage: Same as above except 4-inch size labeled "Standpipe."

2.02 Fire Pump System
  A. General: All equipment and devices shall be U.L. and F.M. approved, and conform to current NFPA Fire Code.
  B. Fire pumps
    (1) Type: Cast iron, bronze trimmed, horizontal split case, double suction, single stage.
    (2) Drive: Open, dripproof induction motor, non-overloading at any point on operating curve. Connect to pump with flexible coupling and provide metal guard over coupling.
    (3) Mounting base: Steel base for pump and motor.
    (4) Manufacturer: Peerless, Worthington, Patterson, Fairbanks-Morse, Aurora.

  C. Fire pump trim
    (1) Compound suction/vacuum gauge, 3½-inch dial spring.
    (2) Discharge pressure gauge, 3½-inch dial spring.
    (3) Automatic air release valve.
    (4) Open screw and yoke gate valve.
    (5) Check valve.
    (6) Casing relief valve.
    (7) Eccentric suction reducer to full pipe size. Concentric discharge increaser to full pipe size.

  D. Fire pump controllers
    (1) Enclosure: Front accessible, floor mounted, drip tight, NEMA II.
    (2) Starter: Part winding for building, across the line for garage, 3 O.L.'s, with manual starter lever.
    (3) Circuit breaker: Externally operated, with time delay trip and instantaneous short circuit trip. Minimum interrupting capacity 22,000 amp. sym. at 480 volts.
    (4) Isolating switch: Externally operated, with latch and latch release to comply with code.
    (5) Other features: Minimum time run relay, "Power On" pilot light, connection for remote "run" pushbutton, power availability signal relay, operating alarm contact, adjustable pressure switch relay.
    (6) Manufacturer: Cutler-Hammer

Bulletin 12030, or approved equal by G.E., Clarke, Lexington, Fairbanks-Morse.
  E. Jockey pumps
     (1) Type: Horizontal centrifugal unit, bronze trim, mounted on cast iron baseplate with open, dripproof induction motor, non-overloading at all points on operating curve.
     (2) Manufacturer: Cutler-Hammer Bulletin 12030, or approved equal by G.E., Clarke, Lexington, Fairbanks-Morse.
  F. Jockey pump controller
     (1) Enclosure: NEMA 3, front accessible.
     (2) Starter: Combination fusible switch, non-reversing magnetic across the line, with Hand-Off Auto selector switch in cover, 3 O.L.'s.
     (3) Other features: Adjustable pressure switch relay, minimum running time relay,
     (4) Manufacturer: Cutler-Hammer Bulletin 12049, or approved equal by G.E., Clarke, Lexington, Fairbanks-Morse.
  G. Auxiliary alarm panel
     (1) Enclosure: Front accessible, wall mounted, with warning lights and pushbutton. Install in Fire Command Room.
     (2) Function: Provide visual and audible signal to indicate
        (a) main pump running
        (b) power to main control panel off
        (c) alarm panel supervisory voltage lost
     (3) Manufacturer: Cutler-Hammer Bulletin 12089H1A, or approved equal by G.E., Clarke, Lexington, Fairbanks-Morse.

2.03 Fire Extinguisher
  A. General: Provide one fire extinguisher for each 2,500 square feet. See plans for specified location. Fire extinguishers in lobby and corridors shall be recessed type, Model AL-215G, finished in dark bronze on an aluminum housing. Parking areas, machine room and electric rooms shall be surface type, Model AL-215-WG, finished same as model above.
  B. Other specified locations
     (a) Two in switch room.
     (b) One in pump room.
     (c) One in the emergency generator room.
     Provide Universal dry chemical type units with 2A-BC rating at all locations, except as noted above.

Part III—Domestic Water System
3.01 General: The hydropneumatic system shall consist of two 300-gallon galvanized tanks; automatic air volume controls; two pump and motor sets; a pre-wired duplex control panel; 1½ HP, 480-volt, 3-phase compressor; 30-gallon tank oil and moisture separator, automatic condensate trap.
  A. Operation
     01 One pump shall be selected as the lead pump and will start and stop in response to its pressure switch which shall be set for 65 psi off, 55 psi on. Should the pressure drop to 50 psi, the second pump (designated as lag pump) shall commence operation and shall also incorporate a minimum run timer which will cause the pumps to operate a minimum of 5 minutes regardless of pressure once they are started.
     02 The air volume controls shall automatically inject air into the tank and maintain proper air/water balance by means of an automatic air release.
  B. Components
     01 Tanks shall be vertical type built for 125 psi working pressure, ASME labeled, hot dip galvanized inside and out after fabrication. Tank shall have required threaded openings for inlet, outlet, drain, and air volume controls and be equipped with a 3/4-inch relief valve set at 125 psi.
     02 Pumps shall be horizontal split case, flexible coupled with mechanical shaft seal.
  C. Acceptable manufacturers: Aurora, Peerless.

Section 17P—Piping Systems

Part I—General
1.01 General Requirements
Construct all piping systems in accordance with applicable codes.

1.02 Utilities
  A. General: Extend all plumbing services to appropriate utility mains and make connections.
  B. Fees and connection charges: Make all necessary arrangements and pay all fees and

connection charges for provision of all required plumbing services to the building and for providing sanitary and storm drainage from the building. This shall include, but not be limited to, provisions for water tap and meter, sanitary service, and storm water drainage.

Part II—Installation Methods

2.01 Excavation

Provide all the excavation and backfilling required for the proper installation of all underground piping and other components installed under grade.

2.02 Install Piping

A. For pipe inside building, install parallel to the lines of the building, close to columns and walls, with vertical pipe truly vertical.

B. To provide adequate access to all equipment and controls.

C. To allow easy draining of water piping, with drain valves at all low points.

D. With all expansion loops, expansion joints, offsets, and anchors required to allow for expansion without damage to the piping or the building, and without objectionable noise.

E. With unions of flanges at each connection to (1) a piece of equipment or control valve; (2) an accessory which requires removal for maintenance.

2.03 Pipe Supports

Support piping must be in accordance with the applicable code. Provide rigid foamglass and galvanized saddle at each support point on insulated piping.

2.04 Install Valves

In accessible locations, and to make possible removal of bonnet and complete servicing of each valve.

2.05 Cleaning Water Piping

Domestic water piping: Flush thoroughly, sterilize with chlorine solution for a minimum of 24 hours, then flush clean. The strength of chlorine solution and methods used must comply with applicable code and health authorities. At completion, there must be no discernible odor. Post warnings until sterilization is complete.

2.06 Testing Piping System

A. General: Test all piping systems to ensure that they are absolutely leak free. The pipe to be insulated shall be proved leak free before any insulation is applied. Wherever pipes will be concealed, a test shall prove the system leak free before the pipe is concealed.

B. Test method: Use the method suitable for the type of piping system being tested. For pressure pipe, use a test pressure approximately 150% of the maximum system working pressure. During this test period, inspect all pipe, fittings, and accessories and eliminate all leaks.

C. Thereafter, where conditions permit, subject each piping system to its normal operating pressure and temperature for not less than 24 hours. The piping systems must remain absolutely tight during this period. The satisfactory completion of any test or series of tests will not relieve the contractor of responsibility for ultimate proper and satisfactory operation of piping systems and their accessories.

2.07 Sleeves

Provide sleeves for all piping passing through masonry walls, floors and roof slabs, except sanitary piping passing through floor slabs on earth fill. Make sleeves of galvanized pipe sized to provide clearance of 1/2 inch around piping, or pipe insulation if the pipe is insulated. Set the end of each sleeve flush with the surrounding surface of wall or ceiling in which the sleeve is installed, but sleeve through floor must project 2 inches above the finished floor. Apply caulk between the sleeve and pipe (or insulation with noncombustible caulking material) as required to maintain a suitable fire and smoke barrier.

2.08 Dielectric Unions

Provide dielectric insulating unions at all connections between dissimilar metals, except at final fixture connections such as galvanized pipe connection to brass water faucet tubing.

Part III—Materials

3.01 Valves

A. Provide valves and accessories of the type and at the locations shown on the drawings or described elsewhere in these specifications. All valves must be produced by a domestic manufacturer.

B. Gate valves 2 inches and smaller: Bronze body, solid wedge, traveling stem, screw-in bonnet, packing held by gland secured by packing nut.

C. Gate valves 2½ inches and larger: Iron body, bronze mounted, outside screw and yoke, rising stem, packing held by gland secured by gland flange.

D. Globe valves 2 inches and smaller: Bronze body, renewable composition disc, slip-on stay-on disc holder, screw over bonnet, packing held by gland secured by packing nut.

E. Globe valves 2½ inches and larger: Iron body, bronze mounted, outside stem and yoke, renewable composition disc and bronze seat ring.

F. Check valves 2 inches and smaller: Bronze body, regrinding swing check, bronze disc.

G. Check valves 2½ inches and larger: Iron body, bronze mounted, regrind-renew swing check, bronze seat ring and disc.

H. Butterfly valves: rated for minimum 125 psi water, with iron body, renewable seat, aluminum bronze or semi-steel disc with welded nickel edge, 316 or 410 stainless steel shaft, extended neck for insulated lines, notched top plate with handle for throttling. Provide a gear operator for valves 6 inches and larger and for valves with chain operators. Valves must not leak at 100 psi across the valves. The valve shall be the full type.

I. Plug valves: Rated for minimum 125 psi water or gas, with semi-steel or iron body, lubricated type 100% pipe area. Provide a wrench operator for valves smaller than 6 inches, a gear operator for valves 6 inches and larger. Valves 2 inches and smaller shall be a screwed type. Valves 2½ inches through 5 inches shall be a flanged type. Valves 6 inches and larger shall be a flanged, gear operated type.

J. Strainers: Iron body, Y pattern or basket type, line size, not more than 2 psi pressure drop, stainless steel screen. Provide a blow-off with gate valve for all strainers.

Part IV—Sanitary Sewer, Outside Building
4.01 General
This paragraph relates to all parts of this system beyond the five foot (more or less in some codes) points outside of building. Extend all sanitary soil and waste drains from building to utility mains or to a location indicated by the architect. Use standard weight, centrifugal cast iron pipe, standard weight fittings, lead or elastomeric joints.

4.02 The following materials are also acceptable, provided they conform to applicable codes.
   A. Schedule 40 PVC pipe and fittings with solvent cemented joints.
   B. Extra strength C-200 vitreous clay pipe and fittings with elastomeric joints.

4.03 Installation and Testing: Continuously grade in the direction of the flow at elevations and slopes as indicated on the drawings. Where elevations and slopes are not indicated, provide maximum grading from connections at the building to the mains, but not in excess of 1/4 inch per foot. Sanitary tees shall not be used in the horizontal position.
   A. Route lines to avoid roots of trees that are to remain.
   B. Test per requirements of the local code and recommendations of the manufacturer.
   C. Cleanouts: Provide cleanouts as indicated on drawings. If not indicated, provide them at points spaced 75 feet apart along the entire length of the drain line and at each change in direction.

Part V—Sanitary Soil, Waste and Vent Piping, Inside
5.01 Building
   A. Standard weight, centrifugal cast iron pipe and standard weight fittings, lead joints, or no-hub cast iron systems and joints, in accordance with the local code.
   B. No-hub cast iron system: Pipe, fittings and couplings shall conform to C.I.S.P.I. Standard No. 301-75.
   C. Vents shall be cast iron as above, or schedule 40 galvanized steel with galvanized malleable iron screwed fittings.
   D. Installation and testing as per the local code.
   E. Flashing: Flash all vent pipes passing through roof with lead roof flashings constructed of number 4 sheet lead, with base extending no less than 10 inches on each side of pipe. The vertical portion of flashing shall extend upward the entire length of the pipe and be turned down inside the pipe at least 2 inches.

Part VI—Exterior Domestic Water
6.01 General: This paragraph relates to all water piping below grade, and all piping above grade outside of building. Extend the water service line, make the connection to the utility main, and provide a meter.

6.02 Materials
   A. Sizes up to 2½ inches: Type K copper rigid water tubing with sweat fittings.
   B. Sizes 3 inches and larger: Centrifugal cast iron, Class 150 pipe with Bell and Spigot push-on joints, Class C cast iron fittings with mechanical joints.

6.03 Meter
   A. Must conform to the requirements of the utility supplying the service.
   B. Shall have the capacity to deliver maximum demand as indicated on drawings with a maximum pressure drop of 5 psi.

C. Provide a stop valve on each side of the meter, and if required by the utility, supply blind flanged valves for a portable bypass, unless required otherwise by the utility:
    (1) Provide oriseal valves for meters 2 inches or smaller.
    (2) Provide U.L. approved O.S.Y. gate valve on the fire meter.

6.04 Meter Box
Provide a meter box and cover as described below, unless the utility supplying the service requires a different installation, in which case, conform to the utility requirements.

6.05 For Meters Larger Than 2 Inches
A. Construct of reinforced concrete, 6-inch walls and bottom, reinforcing both ways all sides and bottom, with a hinged lid marked "water." Must conform to utility department's requirements.
B. Shall be large enough to contain all components and valves, including the transition between different piping materials.
C. Provide sleeves for piping and caulk watertight.

For meters 2 inches and smaller: Use a cast iron meter box with locking cast iron cover marked "water." It shall conform to the utility department's requirements.

Part VII—Interior Domestic Water

7.01 Materials
A. Copper water tube, Type L rigid, with solder type wrought copper fittings, made up with 95-5 solder. Piping under the slabs shall be Type K soft drawn copper installed in PVC sleeves with no joints permitted below the slab. Piping below slabs on grade shall be limited to trap primers only.
B. Valves: Provide a valve in the branch line to each piece of water consuming equipment or fixture. (Stop valves specified under "Section 18P—Plumbing Fixtures" satisfy this requirement.)
C. Air chambers: Install in each water branch, at each fixture, and at each piece of water supplied equipment. Locate in chases or walls as close to the fixture or equipment as possible. Minimum 12-inch high chamber for air.
D. Shock absorbers: Provide factory made shock absorbers immediately adjacent to all equipment and devices which have quick closing valves (except flush valves). Locate them to be easily accessible. They shall be sized, tested, and rated in accordance with Plumbing and Drainage Institute Standards.

7.02 Plumbing Equipment
A. General: Provide all plumbing equipment shown on the drawings, and as specified in this section.
B. Pressure and temperature relief valves
    (1) Provide on water heater and domestic water storage tanks.
    (2) Rating: ASME rated and AGA certified for output of heater, and so labeled.
    (3) Include test lever, extension bulb thermostat and automatic reseating.

7.03 Electric Water Heater
A. General: Constructed in accordance with applicable sections of ASME Code, Certificate of Inspection furnished for minimum 125 psi, ASME labeled, U.L. labeled.
B. Unit shall be insulated with minimum 2-inch thick heavy density glass fiber to conform to the State Energy Code, trimmed with a heavy gauge baked enamel steel jacket, completely factory wired for on point power connection, and 1-inch drain with valve. Provide 3-year tank warranty against leaks from rust or corrosion.

7.04 Components
A. Steel tank with glass lining over base metal and all outlets guaranteed by the manufacturer as suitable for service encountered, to form a non-conducting barrier for tank and system protection.
B. Heating elements constructed of chrome nickel alloy sheath design, selected for maximum of 40 watts per square inch with nickel coating, 4.5 KW per element, <u>2</u> elements. Non-simultaneous operation.
C. Control panel mounted on unit to include magnetic contactors, power circuit fusing 120-volt controls, on-off switch on panel exterior, auxiliary contact for remote on-off control.
D. Two full length magnesium anode rods.
E. Temperature and pressure gauges.
F. High temperature controller.
G. Combination temperature and pressure relief valve, ASME labeled, sized for full input to the unit.
H. Element protection device which de-energizes elements upon low water pressure in the tank.
I. Shunt trip circuit interrupter with externally mounted on-off handle to control all power to the unit. Shall interrupt power upon low water or high temperature conditions.

Capacities: 30-gallon storage, 4.5 KW elements 277 volt, single phase.

Section 18P—Plumbing Fixtures
Part I—General
1.01 Supplementary Documents
Contract Conditions and Division 1 requirements apply to the work specified in this Section.

1.01 General Requirements
    A. Provide all plumbing fixtures and related accessories. The following schedule is based upon model numbers for American Standard and Elkay products. Comparable fixtures as made by Crane, Eljer, Kohler, or Just (unless indicated otherwise) are acceptable provided they meet all requirements of the Contract Documents. Comparable trim as made by Crane, Eljer, Kohler, Just, Chicago, Speakman or T and S (unless indicated otherwise) is acceptable provided it meets all requirements of the Contract Documents.
    B. All plumbing trim shall be provided with renewable seats and replaceable internal working components.
    C. Coordinate exact locations and mounting heights of the fixtures and trim with architect and architectural drawings prior to installation.
    D. All the fixtures, trim, related piping, and accessories shall be rigidly installed and supported.

Part II—Materials and Methods
2.01 Miscellaneous Brass
    A. Traps and drains for sinks: All fixture traps and drains shall be chrome plated 1½-inch, 17-gauge "P" trap with 17-gauge chrome plated tubing to wall and chrome plated escutcheon. Provide cleanouts in the bottoms of all traps.
    B. Traps and drains for lavatories: All fixture traps and drains shall be chrome plated 1¼-inch, 17-gauge "P" trap with 17-gauge chrome plated tubing to the wall and chrome plated escutcheon. Provide cleanouts in the bottom of all traps.
    C. Traps and drains 2 inches and larger for fixtures: All fixture traps and drains 2 inches or larger shall be copper tubing painted aluminum after installation, from the outlet of the fixture to the wall. Provide chrome plated escutcheon at the wall.
    D. Traps and drains for equipment: Same as described above for sinks, lavatories and drains 2 inches and larger. Size is determined by equipment drain connection, but must not be smaller than 1¼ inches unless otherwise indicated on the drawings.
    E. Stops and supplies: Where not specified elsewhere supplies and stops shall be 3/8-inch brass chrome plated American Standard No. 2303.063 loose key type with key and chrome plated escutcheon. The pipe extending from the wall to the stop and from the stop to the fixture connection shall be chrome plated.

2.02 Water Closet, Wall Hung—"WC-1"
    A. Bowl: American Standard No. 2502.011 "Glenco," white, vitreous china, siphon jet, wall hung water closet with elongated bowl and 1½-inch top spud.
    B. Valve: Diaphragm chrome plated quiet flush valve with a 1-inch screwdriver angle stop, a non-hold open feature, a protective cap on the stop, a vacuum breaker and a 1½-inch flush connection. Sloan 115 "Royal" or comparable model by Delany.
    C. Seat: American Standard No. 5320.114 Moltex, heavy duty, white elongated open front seatless cover with stainless steel check hinge or a comparable model by Beneke or Olsonite.
    D. Support: See Section 18P 1.02 D.

2.03 Water Closet, Tank Type—"WC-2"
Same as WC-1 except set carrier at handicapped height.

2.04 Urinal, Siphon Jet, Wall—"U-1"
    A. Urinal: American Standard No. 6540.017 "Allbrook," vitreous china, white, siphon jet, wall hung urinal, flushing rim and 3/4-inch top spud.
    B. Valve: Diaphragm chrome plated quiet flush valve with 3/4-inch screwdriver angle stop, a non-hold open feature, a protective cap and stop, a vacuum breaker and a 3/4-inch flush connection. Sloan 186-11 "Royal" or comparable model by Delany.
    C. Support: See Section 18P 1.02 D.

2.05 Lavatory, Oval Countertop, Hot and Cold Water Spread Fitting—"L-1"
    A. Lavatory: American Standard No. 0470.013 "Ovalyn," 19" x 16", white, vitreous china under counter mounted lavatory with front overflow and rim for under counter mounting with No. 602111-001 mounting frame kit.
    B. Trim: American Standard No. 2238.525 Heritage chrome plated combination faucet with aquaseal valves, aerator, indexed cross

handles, and perforated strainer with 1¼-inch tailpiece.
C. 1/2 G.P.M. hot water flow restrictor and 2.5 G.P.M. total flow restrictor.

2.06 Lavatory, Wall Hung, Hot and Cold Water Spread Fitting—"L-2"
A. Lavatory: American Standard No. 0360.057 "Scotian." 20" x 18", white, vitreous china wall hung lavatory with back, front overflow, soap depression, and concealed arm hanger.
B. Trim: American Standard No. 2238.525 Heritage. Chrome plated combination faucet with aquaseal valves, aerator, indexed cross handles, and perforated strainer with 1¼-inch tailpiece.
C. 1/2 G.P.M. hot water flow restrictor and 2.5 G.P.M. total flow restrictor.
D. Support: See Section 18P 1.02 D.

2.07 Electric Water Cooler, Handicapped, Stainless Steel—"EWC-1"
A. General: A wall-mounted, fully recessed refrigeration unit, 18-gauge type 304 stainless steel, with single bubbler less glass filler, automatic stream regulator with control valve located remote from the bubbler head on the side, 1¼-inch O.D., tailpiece with 1¼-inch cast brass swivel "P" trap with cleanout, stop as specified for lavatory, ARI, U.L. listed, 125 psi working pressure.
B. Refrigeration system: Hermetically sealed 1/5 H.P. compressor, built-in overload, air cooled, pre-cooler, adjustable thermostat control, 5-year warranty.
C. Capacity: 8 gallons per hour based on 80-degree F inlet water and 50-degree F outlet water in 90-degree F room temperature.
D. Mounting: Provide mounting frame and box for in wall installation with all necessary parts and accessories.
E. Manufacturer: Halsey-Taylor model 6800-WCH or comparable model by Elkay, Filtrine, Haws.

Section 18P—Plumbing Drains and Specialties

Part I—General
1.01 Supplementary Documents
Contract Conditions and Division 1 requirements apply to the work specified in this Section.

1.02 General Requirements
A. Provide all drains, accessories and specialties indicated on the Contract Drawings as specified and required under Section 16A, General Instructions, Plumbing Work.
B. Model numbers listed are Josam. Comparable models by Zurn or Smith will be acceptable. Smith only for closet carriers.

Part II—Products and Procedures
2.01 General
A. Provide suitable clamping devices on all drains and cleanouts in the floors with waterproofing membranes.
B. Size of drains shall be indicated on plans. Floor drains shall be a minimum size of 3 inches unless otherwise indicated on Drawings. Outlets shall be either caulked or no-hub as required by the piping used.

2.02 Drains
A. Floor drains
01 Individual showers: Josam 30650-A-3-50 series cast iron two-piece floor drain with seepage flange, reversible clamping collar, positioning set screws, 6-inch diameter Nikaloy strainer, ½-inch trap primer and backwater valve.
02 Toilet rooms, locker rooms, and similar locations: Josam 30650-A-3-50 series cast iron two-piece floor drain with seepage flange, reversible clamping collar, positioning set screws, 6-inch diameter Nikaloy strainer, 1/2-inch trap primer and backwater valve.
03 Equipment rooms and similar areas: Josam 36000-92-50 series cast iron floor drain with flange, seepage opening, 9-inch galvanized grate, clamping device and trap primer.
04 Garage—lower level: Josam 36000-92, 9-inch galvanized grate.
B. Roof drains: Josam 21500 series cast iron roof drain with heavy duty cast iron secured mushroom dome strainer, flashing rim, with integral gravel stop secured with non-corrosive clamping units with the following features as the roof construction required.
01 Extension collar for insulated roof (Series 21000).
02 Bearing pan and underdeck clamp for precast roofs.
03 Parking garage drains (RD-1): Josam 23510-P with heavy duty grate, bearing pan and underdeck clamp.
C. Balcony deck drains (DD-1): Josam 30000A-2 with bearing pan, underdeck clamp, satin bronze top, weepholes.
D. Walkway deck drains (DD-2): Josam 24000-51-1 with deck flange, underdeck clamp satin bronze top.

2.03 Cleanouts

A. General: Install cleanouts at all bends, angles, at ends of all the waste and sewer piping and as noted on Drawings. Bring all cleanouts up to grade on the finished surface and make them easily accessible. All cleanouts shall have extra heavy bronze plugs. For cleanouts in unpaved areas, install in a 18" x 18" x 4" concrete pad with the top flush with finished grade. For cleanouts in paving, sidewalks, etc. install flush with the finished surface.

B. Materials

01 Unfinished areas and chases: Josam 58500 series cast iron caulking ferrule with countersunk, slotted head, lead seal plug.

02 Finished walls: Josam 58740 series cast iron caulking ferrule with countersunk, slotted head, lead seal plug, square smooth finish Nikaloy access cover, 7" x 7" nickel brass frame with anchoring lugs.

03 Finished floors: Josam 58010 series floor cleanout with adjustable extension housing, cast iron ferrule with lead seal, plug, scoriated nickel brass secured access cover and round frame. (When installed in tile floors, the frame shall be square "Type 58030-2." When installed in inlay floors the cover shall be recessed, Suffix "12." When installed in carpeted floors, provide "14" chrome plated carpet marker.)

04 Outside areas: Josam 58090-15 series extra heavy duty cleanout with deep adjustable housing, extra heavy scoriated vandalproof cast iron cover, cast iron ferrule for caulk, plug and positioning set screws. In non-surfaced areas these shall be installed in a 18" x 18" x 4" concrete pad. Locate as directed by the architect.

2.04 Fixture Supports

A. General

01 Provide fixture supports for all lavatories, urinals, and other wall hung fixtures.

02 All foot supports on all types of fixture supports shall be the type that does not extend out from under the wall on which the fixture is mounted.

03 Construction and installation of the supports shall be as required to suit the job conditions, the space available, and the riser diagram and details on the drawings.

B. Carriers

01 Water closet carriers: Josam Unitron adjustable, horizontal or vertical closet carrier as required by waste piping.

02 Urinal carrier: Josam 17810 with bearing plate.

2.05 Hydrants

A. Hose bibbs inside and outside building: Chicago No. 998 rough chrome plated faucet with drain plug, integral vacuum breaker, removable handle, 3/4-inch size.

B. Wall hydrants: Woodford B-74 recessed, polished bronze, loose key, with vacuum breaker.

2.06 Vacuum Breakers

Provide where required by local code, Chicago Faucet Co. No. 892 ½-inch IPS female inlet and outlet, polished chrome plated vacuum breakers.

B

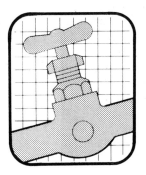

Calculating Water Supply & Fixture Requirements

Section I
Calculating Water Supply Systems

The size of water service and water distributing pipes depends on the type of flush device used on plumbing fixtures, the water pressure, the length of the pipe in the building, the number and kind of

Type of Fixture or Device	Pipe Size (Inches)
Bathtub	½
Combination sink and tray	½
Drinking fountain	⅜
Dishwasher (domestic)	½
Hot water heater	¾ minimum
Kitchen sink, residential	½
Kitchen sink, commercial, one or more compartments	½
Lavatory	½
Laundry tray - 1, 2, or 3-compartment	½
Shower (single head)	½
Sink (service, slop)	½
Sink (flushing rim)	1
Urinal (flush tank)	½
Urinal (direct flush valve)	¾
Water closet (tank type)	½
Water closet (flush valve type)	1
Hose bibbs	½

Size of Fixture Supply Pipe
Table B-1

Fixture	Rate pressure P.S.I.	Flow rate G.P.M.
Ordinary basin faucet	8	3.0
Self-closing basin faucet	12	2.5
Sink faucet - ⅜ inch	10	4.5
Sink faucet - ½ inch	5	4.5
Bathtub faucet	5	6.0
Laundry tub cock - ½ inch	5	5.0
Shower	12	5.0
Ball-cock for water closet	15	3.0
Flush valve for water closet	10-20	15-40
Flush valve for urinal	15	15.0
Garden hose (50 ft.) & sill cock	15	5.0

Rate of Flow and Required Pressure During Flow for Different Fixtures
Table B-2

fixtures installed, and the number of fixtures in use at any given time.

The maximum demand likely to be placed on a building's water supply system can't be determined exactly. You can't predict how many fixtures will be used at the same time. Still, you should estimate the maximum demand so that there will be enough water available to all fixtures when needed. The objective is to avoid undersizing but still to be as economical as possible.

The heart of the water supply system is the water service pipe. It should be sized first. This pipe must

No. of Bathrooms and Kitchens		Diameter of Water Service Pipe	Recommended Meter Size	Approx. Pressure loss Meter and 100' of Pipe	No. of Bathrooms and Kitchens	
Tank-Type Closets					Flush Valve Closets	
Copper	Galvanized	Inches	Inches	P.S.I.	Copper	Galvanized
1-2	--	¾	⅝	27	--	--
--	1-2	¾	⅝	40	--	--
--	--	1	1	30	1	--
3-4	--	1	1	22	--	--
--	3-4	1	1	24	--	--
--	--	1¼	1	32	2-3	--
--	--	1¼	1	36	--	1-2
5-9	--	1¼	1	28	--	--
--	5-8	1¼	1	32	--	--
--	--	1½	1½	29	4-10	--
--	--	1½	1½	30	--	3-7
10-16	--	1½	1½	17	--	--
--	9-14	1½	1½	21	--	--
--	--	2	1½	26	11-18	--
--	--	2	1½	32	--	8-18
17-38	--	2	1½	27	--	--
--	15-38	2	1½	32	--	--
--	--	2	2	25	19-33	--
--	--	2	2	24	--	19-24
39-56	--	2	2	25	--	--
--	39-45	2	2	24	--	--
--	--	2½	2	28	34-57	--
--	--	2½	2	32	--	25-57
57-78	--	2½	2	28	--	--
--	46-78	2½	2	32	--	--
--	--	3	3	16	58-95	--
--	--	3	3	19	--	58-95
79-120	--	3	3	16	--	--
--	79-120	3	3	19	--	--

Residential Use
Minimum Water Service Pipe Size for One and Two-Story Buildings
Hotels, Motels, and Residential Occupancy Only
Table B-3

convey the required volume of water at an acceptable velocity to the water distributing pipes within a building.

The size of the water service pipe must not change when it enters the building and becomes the building water main. This is true whether the pipe is horizontal or vertical (as it would be in a high-rise building). Downstream from where water distributing branch pipes are connected into the building main, the main may be reduced in size with proper fittings. As the main progresses through the building, the demand likely to be placed on the line decreases.

There are two simplified methods for sizing a building's water supply system. Each is based on the expected demand as expressed in units of water supply fixture load. These methods work well for all buildings supplied from a source which has enough water pressure at the highest and most remote fixture during peak demand. Use the tables in this appendix to figure pipe sizes. But keep the following limitations and restrictions in mind:

• The tables apply only where water main pressure does not fall below 50 psi at any time.

• In buildings exceeding two stories but not exceeding four, use the next larger pipe size.

Calculating Water Supply & Fixture Requirements

Number of Fixture Units Flush Tank Water Closet		Size Service	Size Meter	Approximate Pressure Loss	Number of Fixture Units Flush Valve Water Closet	
Copper	Galvanized Iron or Steel	Size Inches	Size Inches	P.S.I.	Copper	Galvanized Iron or Steel
18	--	¾	⅝	30	--	--
--	15	¾	⅝	30	--	--
19-55	--	1	1	30	--	--
--	16-36	1	1	30	--	--
--	--	1	1	30	9	--
56-84	--	1¼	1	30	--	--
--	37-67	1¼	1	30	--	--
--	--	1¼	1	30	10-20	--
--	--	1¼	1	30	--	14
86-255	--	1½	1½	30	--	--
--	68-175	1½	1½	30	--	--
--	--	1½	1½	30	21-77	--
--	--	1½	1½	30	--	15-52
226-350	--	2	1½	30	--	--
--	176-290	2	1½	30	--	--
--	--	2	1½	30	78-175	--
--	--	2	1½	30	--	53-122
351-550	--	2	2	30	--	--
--	291-450	2	2	30	--	--
--	--	2	2	30	176-315	--
--	--	2	2	30	--	123-227
551-640	--	2½	2	30	--	--
--	450-580	2½	2	30	--	--
--	--	2½	2	30	316-392	--
--	--	2½	2	30	--	228-343
641-1340	--	3	3	22	--	--
--	581-1125	3	3	22	--	--
--	--	3	3	22	393-940	--
--	--	3	3	22	--	344-785

Commercial Use
Minimum Water Service Pipe Size for One and Two-Story Buildings
Table B-4

- Buildings located where the water main pressure falls below 50 psi should use the next larger pipe size.
- Table B-1 shows the minimum sizes for fixture supply pipe from the main or from the riser to the wall opening. Follow the listed supply pipe sizes regardless of which method is used.
- Two fixtures is the maximum on a 1/2-inch cold water supply branch.
- Table B-2 gives the required rate of flow and pressure during flow for operating different types of fixtures.

Water supply systems using pipe sizes larger than those included in the two methods illustrated here should be designed by licensed professional engineers.

Method One

Tables B-3 and B-4 are easy to use. First, draw a piping diagram as illustrated in Figure B-5. Hot water piping is not considered when sizing the water service and distribution lines. The horizontal and vertical piping in Figure B-5 serves 18 kitchens and 18 baths in a small apartment building. The street level water pressure is 55 psi. The water piping material selected is galvanized.

Begin with Table B-3, "Residential Use." Start at the most remote outlet "A" and work back

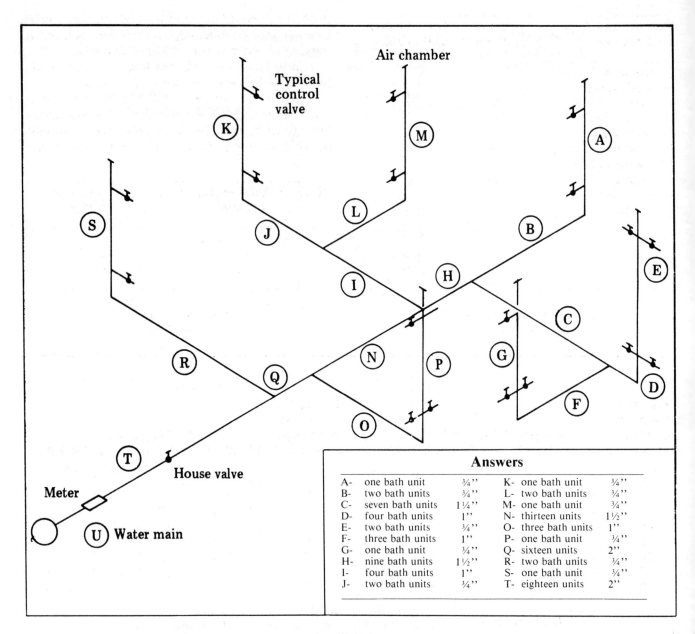

Figure B-5

toward the water main "U." Compute the cumulative demand for the apartment units. Remember that galvanized pipe is being used, so you need to refer to the second column (galvanized) in Table B-3. After you have done your own computing, compare your work with the "Answers" given with the figure.

Method Two
Use the "Factor of Simultaneous Use" table to size water service and distribution lines in a building with fixtures that can't be sized from the number of bathrooms and kitchens (as in Table B-3), or the number and type of water closets (as in Table B-4). The *factor of simultaneous use*, also called the

Number of Fixtures	Percent of Simultaneous Use
1-4	50-100
5-50	25-50
51 or more	10-25

**Factors of Simultaneous Use
Table B-6**

Calculating Water Supply & Fixture Requirements

Number and Type of Fixture	Fixture Units Each	Total Fixture Unit Values
3 - urinal wall type	4	12
6 - flush valve water closets	6	36
1 - service sink	3	3
6 - lavatories	1	6
1 - drinking fountain	½	½
1 - coffee urn ¾" waste	1	1
1 - glass sink 1¼" waste	1	1
		59½ fixture units

Fixture Unit Value as Factor
Table B-7

probable demand on a given installation at a given time, is only an estimate. Table B-6 gives data for making an estimate of probable demand.

To use Table B-6, take the actual number of fixtures installed, not the fixture unit value.

Use a higher number from the "Percentage of Use" ranges for a lower number of fixture ranges. For example, five fixtures would have a probable demand of about 50%, while 45 fixtures would

Fixture Drain or Trap Size in Inches	Fixture Unit Value
1¼ and smaller	1 F.U.
1½	2 F.U.
2	3 F.U.

Special Fixtures
Table B-8

A. ⅜ Inch

Pressure at source in pounds per square inch	Length of Pipe in Feet									
	20	40	60	80	100	120	140	160	180	200
10	5	3	3	2	2	2	--	--	--	--
20	9	5	4	3	3	3	2	2	2	2
30	10	6	5	4	4	3	3	3	3	2
40	--	8	6	5	4	4	4	3	3	3
50	--	9	7	6	5	4	4	3	3	3
60	--	9	7	6	6	5	5	4	4	4
70	--	10	8	7	6	6	5	5	4	4
80	--	--	8	7	7	6	5	5	5	4

B. ½ Inch

Pressure at source in pounds per square inch	Length of Pipe in Feet									
	20	40	60	80	100	120	140	160	180	200
10	10	8	5	5	4	3	3	3	3	3
20	14	10	8	6	6	5	5	4	4	4
30	18	12	10	8	8	7	6	6	5	5
40	20	14	11	10	10	8	7	7	6	6
50	--	16	13	11	11	9	8	7	7	7
60	--	18	14	12	12	10	9	9	8	7
70	--	--	15	13	12	11	10	9	8	8
80	--	--	--	--	--	--	--	--	--	--

C. ¾ Inch

Pressure at source in pounds per square inch	Length of Pipe in Feet									
	20	40	60	80	100	120	140	160	180	200
10	22	14	12	10	8	8	7	6	6	6
20	30	22	18	14	12	12	11	10	10	8
30	38	26	22	18	16	14	13	12	12	10
40	--	30	24	21	19	17	16	16	15	13
50	--	34	28	24	21	19	18	17	16	15
60	--	38	31	26	23	21	20	19	18	17
70	--	--	34	29	25	23	22	21	19	18
80	--	--	36	30	27	24	23	22	21	20

D. 1 Inch

Pressure at source in pounds per square inch	Length of Pipe in Feet									
	20	40	60	80	100	120	140	160	180	200
10	40	28	22	18	16	15	14	13	12	11
20	55	40	32	27	24	22	20	19	18	16
30	70	50	40	34	30	27	25	23	22	20
40	80	58	45	40	35	32	29	27	25	24
50	--	65	57	45	40	36	33	31	29	27
60	--	70	58	50	44	40	36	34	32	30
70	--	76	63	54	45	42	40	37	34	32
80	--	--	65	57	47	43	39	37	35	33

Capacities of Pipe in Gallons per Minute
(Galvanized Iron)
Table B-9

E. 1¼ Inch

Pressure at source in pounds per square inch	\multicolumn{10}{c}{Length of Pipe in Feet}

Pressure at source in pounds per square inch	20	40	60	80	100	120	140	160	180	200
10	80	55	45	37	35	30	27	25	26	24
20	110	80	65	55	50	45	41	38	36	34
30	--	100	80	70	60	56	51	47	45	42
40	--	--	95	80	72	65	60	56	52	50
50	--	--	107	92	82	74	68	63	60	55
60	--	--	--	102	90	81	75	70	65	62
70	--	--	--	--	97	88	82	74	69	67
80	--	--	--	--	105	95	87	79	74	72

F. 1½ Inch

Pressure at source in pounds per square inch	20	40	60	80	100	120	140	160	180	200
10	120	90	70	60	55	50	45	40	40	35
20	170	130	100	90	75	70	65	60	55	55
30	--	160	130	110	100	90	80	75	70	65
40	--	170	150	130	110	100	90	90	80	80
50	--	--	170	140	130	120	110	100	90	90
60	--	--	--	160	140	130	120	110	100	100
70	--	--	--	170	150	140	130	120	110	100
80	--	--	--	--	160	150	140	130	120	110

G. 2 Inch

Pressure at source in pounds per square inch	20	40	60	80	100	120	140	160	180	200
10	240	160	130	110	100	90	90	80	80	70
20	300	240	200	160	150	140	130	120	110	100
30	--	300	240	200	180	160	150	140	140	130
40	--	--	280	240	220	200	180	160	160	150
50	--	--	--	280	240	220	200	200	180	160
60	--	--	--	--	280	240	220	200	200	180
70	--	--	--	--	300	260	240	220	220	200
80	--	--	--	--	--	280	260	240	220	220

**Capacities of Pipe in Gallons per Minute
(Galvanized Iron)
Table B-9 (continued)**

have a probable demand of about 25%. If a table showing factors of simultaneous use is not available, a practical way of estimating the probable demand is 30% of the maximum fixture demand in gallons. The demand values assigned to fixtures are the same for sizing water supply pipes as for sizing drainage, waste, and vent pipes. One unit equals 7½ gallons per minute. (See "Fixture Unit Value Table" B-7 and "Special Fixtures Table" B-8.)

Use Tables B-9 and B-10 for galvanized pipe and copper tubing. Calculate the maximum fixture demand and figure the factor of simultaneous use to find the correct size of pipe for the water service line. The minimum size for a water service line is 3/4 inch. *Use 3/4-inch pipe even when calculations show a smaller one could be installed.*

The maximum fixture demand in gallons is based on the number and type of all fixtures in the plumbing system. Use Tables B-7 and B-8 to find the maximum fixture demand. For example, assume a plumbing system with 3 urinals, 6 flush valve water closets, one service sink, 6 lavatories, 1 drinking fountain, a coffee urn with 3/4-inch waste, and a glass sink with 1¼-inch waste.

Use Tables B-7 and B-8 to find the maximum fixture demand of **446** gallons per minute. All fixtures would never be used at the same time, so the maximum fixture demand is reduced by the factor of simultaneous use found in Table B-6.

Calculating Water Supply & Fixture Requirements

A. ½ Inch

Pressure at source in pounds per square inch	\multicolumn{10}{c}{Length of Pipe in Feet}

Pressure at source in pounds per square inch	20	40	60	80	100	120	140	160	180	200
10	8	5	4	3	3	2	2	2	2	2
20	12	8	6	5	5	4	4	3	3	3
30	15	10	8	7	6	5	5	4	4	4
40	17	12	9	8	7	6	6	5	5	4
50	--	14	10	9	8	7	6	6	5	5
60	--	15	12	10	9	8	7	7	6	6
70	--	--	13	11	10	9	8	7	7	6
80	--	--	14	12	10	10	8	8	7	7

B. ⅝ Inch

Pressure at source in pounds per square inch	20	40	60	80	100	120	140	160	180	200
10	12	8	7	6	5	5	4	4	3	3
20	18	12	10	9	7	6	6	5	5	5
30	22	16	12	10	9	9	8	7	6	6
40	26	18	14	12	10	10	9	8	8	7
50	--	22	16	14	12	11	10	9	9	8
60	--	24	18	16	14	13	12	11	10	9
70	--	--	20	18	15	14	13	12	11	10
80	--	--	22	19	16	15	14	13	12	11

C. ¾ Inch

Pressure at source in pounds per square inch	20	40	60	80	100	120	140	160	180	200
10	20	14	10	10	8	8	6	6	6	5
20	30	20	16	14	12	10	10	10	8	8
30	36	26	20	17	15	14	13	11	10	8
40	--	30	24	20	18	16	15	14	13	12
50	--	34	28	24	20	18	16	16	14	14
60	--	36	30	26	22	20	18	18	16	16
70	--	--	32	28	24	22	20	18	18	16
80	--	--	36	30	26	24	22	20	18	18

D. 1 Inch

Pressure at source in pounds per square inch	20	40	60	80	100	120	140	160	180	200
10	50	30	24	20	18	16	14	14	12	12
20	70	45	36	30	26	24	22	20	18	18
30	80	55	45	38	34	30	28	26	24	22
40	--	65	55	45	40	36	32	30	28	26
50	--	75	60	50	45	40	36	34	32	30
60	--	80	65	55	50	45	40	38	36	34
70	--	--	70	60	55	50	45	40	38	36
80	--	--	80	65	60	50	50	45	40	40

E. 1¼ Inch

Pressure at source in pounds per square inch	20	40	60	80	100	120	140	160	180	200
10	80	55	42	37	32	30	27	25	22	22
20	110	80	65	55	47	42	40	35	35	32
30	---	105	80	70	60	55	50	45	42	40
40	---	110	95	80	70	65	60	55	50	47
50	---	---	110	90	80	70	65	60	57	55
60	---	---	---	105	90	80	75	70	65	60
70	---	---	---	110	100	90	80	75	70	65
80	---	---	---	---	105	95	85	80	75	70

F. 1½ Inch

Pressure at source in pounds per square inch	20	40	60	80	100	120	140	160	180	200
10	130	90	70	60	50	45	40	40	35	35
20	170	130	100	90	75	70	65	60	55	50
30	---	170	130	110	100	90	80	75	70	65
40	---	---	155	130	115	105	95	88	80	77
50	---	---	170	150	130	120	108	100	90	88
60	---	---	---	165	145	130	120	110	105	98
70	---	---	---	170	160	142	130	122	113	106
80	---	---	---	---	170	155	140	130	122	115

**Capacities of Pipe in Gallons per Minute
(Copper Tubing)
Table B-10**

G. 2 Inch

Pressure at source in pounds per square inch	Length of Pipe in Feet									
	20	40	60	80	100	120	140	160	180	200
10	280	180	150	145	110	100	90	85	80	70
20	320	280	220	190	165	160	140	125	120	110
30	---	320	280	240	210	180	170	160	150	140
40	---	---	320	280	240	220	200	190	175	160
50	---	---	---	320	280	250	230	210	200	190
60	---	---	---	---	300	280	260	240	220	200
70	---	---	---	---	320	300	280	260	240	230
80	---	---	---	---	---	320	300	280	260	240

Capacities of Pipe in Gallons per Minute
(Copper Tubing)
Table B-10 (continued)

The 19 fixtures would have a factor of simultaneous use of about 35%. The maximum fixture demand was 446 gallons per minute. Thus the water service line must have a capacity of 35% of 446 or 156 gallons per minute. Assume pipe 60 feet long and a pressure at the main of 50 psi. Tables B-9 and B-10 show that either a 1½-inch galvanized pipe or a 1½-inch copper tubing water service line would be needed.

Section II
Calculating Fixture Requirements and Sewage Flow

Architects may occasionally ask you to help design a building drainage system. You should be able to calculate the number of people who will use the system. Then the architect can determine the number and type of fixtures required and the sewage flow. The drainage, waste, vent and water piping are then sized accordingly.

Of course, after you have done this work for the architect, you expect to do the installation. And in most cases you will. It's good business to work closely with architects when your services are requested.

The following illustrations and tables will help you determine the number and types of fixtures for various types of occupancies. The examples and illustrations in this section deal exclusively with commercial buildings. These are more complex than the simple residential jobs covered so far in this manual.

Places of Employment

The number of toilet fixtures in manufacturing plants, warehouses and similar establishments is based on the assumed number of employees. The percentage ratio and type of fixtures required for both sexes may be changed by the administrative authority. The authority in your area may be willing to alter the requirements in Table B-11, "Places of Employment," if you can provide information which shows that some other fixture ratio is more appropriate.

Let's consider one such example. Assume that toilet facilities are needed for a medium-size manufacturing plant employing 100 persons. Some building codes require a ratio of 50% male and 50% female facilities. Other codes for the same type of occupancy use a percentage of 75% to 25%. Obviously, if these ratios were used rigidly, an imbalance of toilet fixtures would result in many installations. The administrative authority (plumbing plans examiner) may request a letter from the owner giving the maximum number of probable male and female employees in his particular plant. Then the correct number and type of plumbing fixtures is figured from Table B-11.

Additional requirements. A drinking fountain must be provided for each 75 persons or portion thereof. It must be located within 50 feet of all operations. Drinking fountains cannot be located in any restroom or vestibule to a restroom.

Wash-up sinks may be substituted for lavatories

Calculating Water Supply & Fixture Requirements

| | **Males** | | | | **Females** | |
No. of Males	Water Closets	Urinals	Lavatories	No. of Females	Water Closets	Lavatories
10- 30	1	1	2	1- 12	1	2
31- 46	2	1	3	13- 34	2	3
47- 63	2	2	4	35- 58	3	4
64- 80	3	2	5	59- 83	4	5
81- 96	3	3	6	84-109	5	6
97-116	4	3	7	110-138	6	8
117-136	5	3	8	139-170	7	9
137-156	5	4	9	171-200	8	11
157-177	6	4	10	--	--	--
178-200	7	4	11	--	--	--

For each group of 25 males over 200, add one water closet, one urinal and one lavatory.

For each group of 30 females over 200, add one water closet and one lavatory. Female urinals may be substituted for water closets but the number can not exceed one-half of the number of water closets required. In other words, if six water closets are required, either one, two or three of the six water closets required could be female urinals.

Place of Employment
Table B-11

where the type of employment warrants their use.

Establishments that have 10 or more offices or rooms and employ 25 persons or more must provide a service sink on each floor.

Manufacturing plants that may subject employees to excessive heat, infection or irritating materials must provide a shower for each 15 persons.

Establishments employing 9 persons or less which do not cater to the public (such as storage warehouses and light manufacturing buildings) have less rigid requirements. In these applications some codes would consider as adequate one water closet and one lavatory for both sexes. But the following conditions must be observed:

• If the minority sex exceeds 3 persons, separate toilet facilities are required. For example, where 4 males and 5 females (or vice versa) are employed, separate toilet facilities must be provided.

• If the number of males employed exceeds 5, a urinal must be provided.

Public Assembly

Establishments which are used by the public must have toilet facilities for the number of employees and the public reasonably anticipated, unless they have special permission to do otherwise. There are two classifications of public use which determine the number and type of plumbing fixtures required:

1. Establishments that provide countable seating capacity such as churches, theaters, stadiums and restaurants.

2. Establishments such as retail stores, office buildings, and the like that have no countable seating capacity. The facilities required are determined by the square foot area.

Public places having seating capacities. Public assembly facilities such as churches, theaters, stadiums and similar establishments use the percentage ratio in determining the number and type of plumbing fixtures required. Codes vary considerably in the assumed ratio of male and female. So it's important to know the ratio used in your local code.

Once the number of males and females is established, determine the number of fixtures required from Table B-12, "Public Assembly."

The number and type of plumbing fixtures required for food and drink establishments is determined, in large part, by whether or not those establishments serve alcoholic beverages. A ratio of 50% male and 50% female is usually assumed.

First, find the number of occupants by adding the seating capacity. Then use Table B-13 or Table

| | Males | | | | Females | |
No. of Males	Water Closets	Urinals	Lavatories	No. of Females	Water Closets	Lavatories
1- 100	1	1	1	1- 50	1	1
101- 250	2	1	1	51- 140	2	1
251- 360	2	2	1	141- 250	3	2
361- 470	2	3	2	251- 360	4	2
471- 580	3	3	2	361- 470	5	3
581- 700	3	4	3	471- 690	6	3
701- 820	3	5	3	691- 960	7	4
821- 975	4	5	4	961-1300	8	4
976-1150	4	6	4	1301-1640	9	5
1151-1325	4	7	4	1641-2000	10	6
1326-1490	5	7	5	2001-2350	11	7
1491-1675	5	8	5	2351-2700	12	8
1676-1875	5	9	5	--	--	--
1876-2075	6	9	6	--	--	--
2076-2250	6	10	6	--	--	--
2251-2475	6	11	6	--	--	--
2476-2700	6	12	7	--	--	--

For each group of 500 males over 2700, add one water closet, and one lavatory. For each additional 300 males, add one urinal.

For each group of 350 females over 2700, add one water closet. For each additional group of 500 females, add one lavatory.

Drinking fountains must be provided at a ratio of one for each 200 persons.

Public Assembly
Table B-12

| | Males | | | | Females | |
No. of Males	Water Closets	Urinals	Lavatories	No. of Females	Water Closets	Lavatories
1- 62	1	1	1	1- 30	1	1
63- 98	2	1	1	31- 62	2	1
99-138	2	2	2	63- 98	3	1
139-181	2	3	2	99-138	4	2
182-226	3	3	2	139-181	5	2
227-272	3	4	3	182-226	6	2
273-320	3	5	3	227-272	7	8
321-369	4	5	3	273-320	8	8
370-420	4	6	4	321-369	9	8
--	--	--	--	370-420	10	4

Food and Drink Establishments Serving No Alcohol
Table B-13

B-14 to select the type and number of fixtures required.

Public places without seating capacities. Public places such as shopping centers, retail stores, office buildings and similar establishments use the square foot of floor method to determine the rated capacity. The net and gross square footage per expected

Calculating Water Supply & Fixture Requirements

| | Males | | | | Females | |
No. of Males	Water Closets	Urinals	Lavatories	No. of Females	Water Closets	Lavatories
1- 42	1	1	1	1- 20	1	1
43- 65	2	1	1	21- 42	2	1
66- 90	2	2	2	43- 65	3	1
91-117	2	3	2	66- 90	4	2
118-147	3	3	2	91-117	5	2
148-178	3	4	3	118-147	6	2
179-212	3	5	3	148-178	7	3
213-247	4	5	3	179-212	8	3
248-282	4	6	4	213-247	9	3
283-317	4	7	4	248-282	10	4
318-352	5	7	5	283-317	12	4
353-390	5	8	5	318-352	12	5
--	--	--	--	353-390	13	5

Food and Drink Establishments Serving Alcoholic Beverages
Table B-14

Type of occupancy	Square feet per occupant	Type of occupancy	Square feet per occupant
Aircraft hangars (no repair)	500 gross	Hospitals. sanitariums and nursing homes, sleeping areas	120 gross
Auction rooms, assembly areas, concentrated use (without fixed seats), auditoriums, bowling alleys, (assembly areas), dance floors, lodge rooms, reviewing stands, stadiums	7 net	Institutional areas	240 gross
		Hotels and apartments	200 gross
		Industrial and manufacturing	100 gross
		Kitchens, commercial	200 gross
		Library reading rooms	100 gross
Assembly areas, less concentrated use, conference rooms, dining rooms, drinking establishments, exhibit rooms, gymnasiums, lounges, skating rinks, stages	15 net	Toilets and locker rooms	50 gross
		Mechanical equipment rooms	300 gross
		Nurseries for children (day care)	20 net
		Offices	100 gross
Assembly areas, standing or waiting spaces	3 net	School shops and vocational rooms	50 net
		Storage, shipping and similar uses	100 gross
Children's homes and homes for the aged	120 gross	Storage warehouses	1,500 gross
		Stores, retail sales rooms	
Classrooms	20 net	Basement	30 gross
Dormitories	200 gross	Ground floor	30 gross
Garage parking	200 gross	Upper floors	60 gross

Footage per Expected Occupant
Table B-15

Types of Establishments	Gallons Per Day (GPD)
Residential	
Single family	350 GPD
Townhouse	150 GPD
Apartment	150 GPD
Mobile home	350 GPD
Duplex or twin home	350 GPD*

*Same as single family, for entire structure

Commercial	
Barber shop	200 GPD/chair
Beauty salon	200 GPD/chair
Bowling alleys	100 GPD/lane
Dentist offices a) per dentist	375 GPD/Dr.
Dentist offices b) per wet chair	200 GPD/chair
Doctor offices (per doctor)	375 GPD/Dr.
Full service restaurant	50 GPD/seat
Bar and lounge	15 GPD/seat
Fast food restaurant	35 GPD/seat
Take-out restaurant	50/100 GPD/sq. ft.
Hotels and motels	200 GPD/room
Laundries, self-service	400 GPD/washer
Office building	20/100 GPD/sq. ft.
Service stations	20/100 GPD/sq. ft.
Shopping centers	20/100 GPD/sq. ft.
Stadiums, race tracks, ballparks	3 GPD/seat
Stores, without food service	20/100 GPD/sq. ft.
Theaters a) indoor, auditoriums	3 GPD/sq. ft.
Theaters b) outdoor, drive-ins	5 GPD/car space
Camper or trailer parks	200 GPD/space
Industrial	
Factories (exclusive of industrial waste) without showers	20/100 GPD/sq. ft.
With showers	35/100 GPD/sq. ft.
Airports	5 GPD/passengers
and	20 GPD/employees
Churches	3 GPD/seat
Hospitals	250 GPD/bed
Nursing, rest homes	150 GPD/bed
Parks, public picnic	
a) with toilets only	5 GPD/person
b) with bath house, showers and toilets	10 GPD/person
Public institutions other than hospitals	
a) jail, boarding school, etc.	125 GPD/person
b) school	15 GPD/person
c) swimming and bathing facilities, public	10 GPD/person
Warehouse/industrial-speculation building	5/1000 GPD/sq. ft.
Storage warehouse or mini-warehouse	20/1000 GPD/sq. ft.

Daily Sewage Flow
Table B-16

occupant is given in Table B-15. Divide the net or gross square feet of occupancy into the gross square feet of floor space for the type of building involved. This gives you the number of persons that can legally occupy a building. Check the applicable code for the assumed ratio of male and female. Use the appropriate table to determine the number and type plumbing fixtures required.

Sewage Flow

Sometimes you may have to size drainage, waste, vent and water piping systems when the type of occupancy is known but the net or gross square footage is unknown. Table B-16 gives the estimated average daily sewage flow for various types of establishments.

C

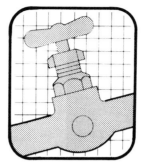

Making Plumbing Calculations

Every plumbing estimator should have a good working knowledge of the mathematics needed for making plumbing calculations. For example, you should be able to figure the area of a pipe and the contents of a cylindrical tank. The following tables will be useful references:

Circumference	= diameter x 3.1416
Circumference	= radius x 6.2832
Circumference	= square root of area x 3.54
Diameter	= circumference x 0.3183
Diameter	= square root of area x 1.1283
Radius	= circumference x 0.1591
Radius	= square root of area x 0.564
Area	= diameter squared x 0.7854
Area	= radius squared x 3.1416
Area	= circumference squared x 0.0796
Area	= circumference x diameter x 0.25

Equations Referring to a Circle
Table C-1

Circles

The *diameter* of a circle is a straight line passing through the center and terminating at each edge.

The *radius* of a circle is a straight line extending from the center to the edge. In other words, it is one-half the diameter.

The *circumference* of a circle is the distance around the outer edge of a circle.

Pipe Size in Inches	Circumference	Pipe Size in Inches	Circumference
⅛	.3927	2	6.2832
¼	.7854	2½	7.8540
⅜	1.1781	3	9.4248
½	1.5708	3½	10.9956
¾	2.3562	4	12.5664
1	3.1416	5	15.7080
1¼	3.9270	6	18.850
1½	4.7124	8	25.133

Circumferences of Standard Size Pipes
Table C-2

Pipe Size in Inches	Area	Pipe Size in Inches	Area
⅛	0.0123	2	3.141
¼	0.0491	2½	4.91
⅜	0.1104	3	7.068
½	0.1963	3½	9.621
¾	0.4417	4	12.566
1	0.7854	5	19.635
1¼	1.227	6	28.274
1½	1.767	8	50.265

Area of Standard Sized Pipe Openings
Table C-3

Diameter	¼	⅜	½	¾	1	1¼	1½	2	3	4	6
¼	1	2.7	5.6	15.6	32.3	56.3	--	--	--	--	--
⅜	--	1	2	5.6	11.6	20.3	32	65.7	--	--	--
½	--	--	1	2.7	5.6	9.8	15.5	32	--	--	--
¾	--	--	--	1	2	3.8	5.6	11.6	32	--	--
1	--	--	--	--	1	1.7	2.7	5.6	15.5	32	88.1
1¼	--	--	--	--	--	1	1.5	3.2	8.9	18.3	50.4
1½	--	--	--	--	--	--	1	2	5.6	11.6	32
2	--	--	--	--	--	--	--	1	2.74	5.6	15.5
3	--	--	--	--	--	--	--	--	1	2	5.6
4	--	--	--	--	--	--	--	--	--	1	2.7
6	--	--	--	--	--	--	--	--	--	--	1

Pipe Equivalents
Table C-4

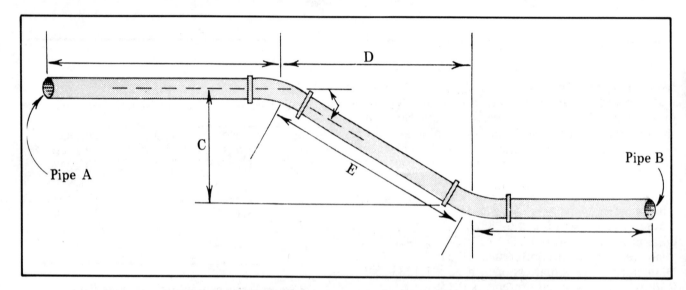

Common Pipe Offset Between Parallel Runs of Pipe
Figure C-5

Degree of Offset	When C = 1, D =	When D = 1, C =	When C = 1, E =
60°	0.5773	1.732	1.1547
45°	1.000	1.000	1.4142
30°	1.732	0.5773	2.000
22½°	2.414	0.4142	2.6131
11¼°	5.027	0.1989	5.1258
5⅝°	10.168	0.0983	10.217

Finding the Length of Pipe to Connect Two Parallel Runs of Piping
Table C-6

Making Plumbing Calculations

Dia. in Feet	Dia. in Inches	U.S. Gallons	Dia. in Feet	Dia. in Inches	U.S. Gallons
1	0	5.86	3	10	86.32
1	2	8.00	4	0	94.00
1	4	10.44	4	2	102.0
1	6	13.22	4	4	110.2
1	8	16.32	4	6	119.0
1	10	19.76	4	8	128.0
2	0	23.50	4	10	137.2
2	2	29.57	5	0	147.0
2	4	32.00	5	2	156.7
2	6	36.73	5	4	167.0
2	8	41.78	5	6	177.6
2	10	47.15	5	8	188.6
3	0	52.89	5	10	200.0
3	2	58.91	6	0	211.4
3	4	65.28	6	6	248.2
3	6	71.98	7	0	288.0
3	8	79.00	7	6	330.4

Volume of Cylindrical Tanks in U.S. Gallons per Foot of Depth
Table C-7

Width Feet	Length in Feet - of Tank						
	2	2½	3	3½	4	4½	5
2	29.92	37.40	44.88	52.36	59.84	67.32	74.81
2½	--	46.75	56.10	65.45	74.81	84.16	93.51
3	--	--	67.32	78.55	89.77	101.0	112.2
3½	--	--	--	91.64	104.7	117.8	130.9
4	--	--	--	--	119.7	134.6	149.6
4½	--	--	--	--	--	151.5	168.3
5	--	--	--	--	--	--	187.0

Volume - Rectangular Tanks, Capacity in U.S. Gallons per Foot of Depth
Table C-8

For practice: The cross section area of a 2-inch pipe is equivalent to the cross section area of how many 1-inch pipes?

Answer: Look in Table C-4 in the vertical column under 2-inch, to where it meets the horizontal line for 1-inch (bold type). This tells you that it requires 5.6 one-inch pipes to equal the cross sectional area of one 2-inch pipe.

Measuring Offsets

Laying out and dimensioning piping arrangements can involve complex calculations of pipe length. The only place offsets are not a problem is where 90-degree elbows are used. But you can find the exact distance between centers of fittings on offsets by careful calculation.

Figure C-5 shows a common pipe offset between parallel runs of pipe. The center to center offset (C) is 10 inches, the distance between the centers of parallels A and B. Assume that pipes A and B are to be connected with 45-degree elbows. You need to determine the length of pipe required for E.

Use Table C-6 which refers to the letters on Figure C-5. Find the number 1.4142 opposite the 45-degree offset. Multiply this figure by distance C, in this case 10 inches. The length of pipe required would be 14.14 inches.

Assume you have to find the capacity in U.S. gallons of a rectangular tank. If the dimensions are in feet, use the following procedure: multiply the three dimensions, (length x width x depth) by 7.476. This yields the capacity in U.S. gallons. If the dimensions are in inches, use the following pro-

1 cubic foot of water weighs	62.5 pounds
1 cubic inch of water weighs	.03617 pounds
1 cubic inch of water contains	.004329 gallons
Gallons x .16	= cubic feet
Cubic inches ÷ 231	= gallons
Cubic inches x .0043	= gallons
1 gallon water	= 8.338 pounds
1 gallon water	= 231 cubic inches
1 cubic foot water	= 7.476 gallons
1 pound water	= 27.7 cubic inches

**Hydrostatic Table
Table C-9**

Pressure in Pounds per Square Inch at Various Head Elevations of Water in Feet

Head Feet	Pressure in Pounds Per Square Inch
0	---
1	.43
10	4.34
20	8.67
30	13.01
40	17.34
50	21.68
60	26.01
70	30.35
80	34.68
90	39.02
100	43.35

**Pressure in Pounds Per Square inch = the Head in Feet x .4335
Table C-10**

Heads of Water in Feet Showing Pressures in Pounds per Square Inch

Head Feet	Pressure in Pounds Per Square Inch
0	---
1	2.31
10	23.10
20	46.20
30	69.30
40	92.40
50	115.50
60	138.60
70	161.70
80	184.80
90	207.90
100	231.00

**Head of Water in Feet x 2.31 = Pressure in Pounds per Square Inch
Table C-11**

Fixture	Flow Pressure PSI	Demand GPM
Ordinary lavatory faucet	8	2.5
Self-closing lavatory faucet	12	3.0
Sink faucet, ⅜ or ½ inch	6	4.5
Sink faucet, ¾ inch	7	6.0
Bathtub diverter valve	5	5.0
Shower valve	12	5.0
Laundry tub faucet, ½ inch	5	5.0
Ballcock type water closet	15	3.0
Flush valve type water closet	10-20	35.0
Flush valve type urinal	15	27.0
Drinking fountain (average)	5	0.8
¾ inch, garden hose sill cock	30	5.0

**Rate of Flow and Required Pressure at Individual Water Outlets
Table C-12**

cedure: multiply the three dimensions, (length x width x depth). Then divide the result by 231. The answer is the capacity in U.S. gallons.

Flow pressure is the pressure in pipe at the water outlet which serves a fixture.

The figures given in Table C-12 are based on water conservation measures now in effect. Many model codes are considering other flow rates. Should these be changed, certain fixture water

Making Plumbing Calculations

outlets may be reduced slightly below those shown in Table C-12 without causing any noticeably adverse effect in fixture use.

It's difficult to generalize about flow pressure and gallonage required for flush-valve type water closets. There are too many designs and types of valves.

The minimum size of a fixture-supply pipe from the riser or main to the wall opening should be as listed in the following table. Most codes accept these values. (See Table C-13.)

For fixtures not listed, the minimum supply pipe size may be the same as for a comparable fixture.

Some codes do not permit more than two fixtures to be connected to a 1/2-inch cold water supply branch.

Type of Fixture or Device	Pipe Size (Inches)
Bathtubs	½
Combination sink and tray	½
Drinking fountain	⅜
Dishwasher (domestic)	½
Hot water heaters (minimum cold water)	¾
Kitchen sink, residential	½
Kitchen sink, commercial	¾
Lavatory	½
Laundry tray 1, 2 or 3 compartments	½
Shower (single head)	½
Sinks (service type)	½
Sinks with flushing rim	1
Urinal with flush tank	½
Urinal with direct flush valve	¾
Water closet tank type	½
Water closet with direct flush valve	1
Hose bibbs	½ or ¾

**Pipe Sizes for Various Fixtures
Table C-13**

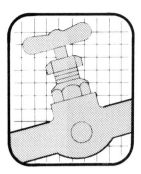

Definitions

Definitions

The terms included here are found in most plumbing codes. Some words in the code have become so descriptive and specialized that their meaning is different from what a dictionary might give.

Two words that appear repeatedly in every code are *shall* and *may*. To be able to comply with the code, you must clearly understand the specialized meanings of these words.

Shall means that compliance is mandatory and that the procedure or condition specified must be performed without deviation. For example, part of the requirements for waste disposal is that "sewage and liquid waste *shall* be treated and disposed of as hereinafter provided." *May* is a term of permission. When used in the code, it means "allowable" or "optional," but not required. For example, "Drinking fountains *may* be installed with indirect waste only for the purpose of resealing required traps of floor drains."

"Building drain" and "building sewer" are two terms often used improperly. Many professionals assume that both terms apply to the same part of the drainage system. However, note the code definition for each term: Building drain is the main horizontal collection system within the walls of a building which extends to five feet beyond the building line. (This distance may vary in some codes.) A building sewer is defined as that part of the horizontal drainage system outside the building line which connects to the building drain and conveys the liquid waste to a legal point of disposal.

Effective and constructive code interpretation is possible only when the words and terms used in the code are understood. Many definitions can be illustrated through isometric drawings. It is possible to have some variation and still remain within the intent of the code definitions. The alert professional will discover that most isometric drawings fit into these definitions. Isometric illustrations provide a better understanding of code definitions.

Absorption Drainfield absorption area.

Air gap (in a water supply system) The unobstructed vertical distance through the free atmosphere between the lowest opening from any pipe or faucet supplying water to a tank, plumbing fixture, or other device, and the flood level rim of the receptacle.

Anaerobic Living without free oxygen. Anaerobic bacteria found in septic tanks are beneficial in digesting organic matter.

Approved Approved by the plumbing official or other authority given jurisdiction by the code.

Area drain A receptacle designed to collect surface or rain water from an open area.

Backfill That portion of the trench excavation up to the original earth line which is replaced after the sewer or other piping has been laid.

Backflow The flow of water or other liquids, mixtures, or substances into the distributing pipes of a potable supply of water and any other fixture or appliance, from any source, which is opposite to the intended direction of flow.

Backflow connection Any arrangement whereby backflow can occur.

Backflow preventer A device or means to prevent backflow into the potable water system.

Back Siphonage The flow of water or other liquids, mixtures or substances into the distributing pipes of a potable supply of water or any other fixture, device, or appliance, from any source, which is opposite to the intended direction of flow due to negative pressure in the pipe.

Base The lowest point of any vertical pipe.

Battery of fixtures Any group of two or more similar adjacent plumbing fixtures which discharge into a common horizontal waste or soil branch.

Boiler blow-off An outlet on a boiler to permit emptying or discharging of water or sediment in the boiler.

Branch Any part of the piping system other than a main, riser or stack.

Branch interval A length of soil or waste stack (vertical pipe) generally one story in height (approximately nine feet, but not less than eight feet) into which the horizontal branches from one floor or story of a building are connected to the stack.

Branch vent The vent that connects one or more individual vents with a vent stack or stack vent.

Building drain The main horizontal sanitary collection system, inside the wall line of the building, which conveys sewage to the building sewer beginning five feet (more or less in some codes) outside the building wall. The building drain excludes the waste and vent stacks which receive the discharge from soil, waste and other drainage pipes, including storm water.

Building sewer That part of the horizontal piping of a drainage system which connects to the end of the building drain and conveys the contents to a public sewer, private sewer, or individual sewage disposal system.

Building storm drain A drain used to receive and convey rain water, surface water, ground water, subsurface water and other clear water waste, and discharge these waste products into a building storm sewer or a combined building sewer beginning five feet outside the building wall.

Building storm sewer Connects to the end of the building storm drain to receive and convey the contents to a public storm sewer, combined sewer, or other approved point of disposal.

Building subdrain Any portion of a drainage system which cannot drain by gravity into the building sewer.

Caulking Any approved method of rendering a joint water and gas tight. For cast iron pipe and fittings with hub joints, the term refers to caulking the joint with lead and oakum.

Code Regulations and their subsequent amendments or any emergency rule or regulation lawfully adopted to control the plumbing work by the administrative authority having jurisdiction.

Combined building sewer A building sewer which receives storm water, sewage and other liquid waste.

Common vent The vertical vent portion serving to vent two fixture drains which are installed at the same level in a vertical stack.

Conductor See Leader.

Continuous waste A drain connecting a single fixture with more than one compartment or other permitted fixtures to a common trap.

Cross connection Any physical connection or arrangement between two separate piping systems, one containing potable water and the other water of unknown or questionable safety.

Dead end A branch leading from a soil, waste or vent pipe, building drain or building sewer which is terminated by a plug or other closed fitting at a developed distance of two feet or more. A dead end is also classified as an extension for future connection, or as an extension of a cleanout for accessibility.

Developed length The length as measured along the center line of the pipe and fittings.

Diameter The nominal diameter of a pipe or fitting as designed commercially, unless specifically stated otherwise.

Downspout See Leader.

Drain Any pipe which carries liquid, waste water or other water-borne wastes in a building drainage system to an approved point of disposal.

Drainage system All the piping within a public or a private premises that conveys sewage, rain water, or other types of liquid wastes to a legal point of disposal.

Drainage well Any drilled, driven or natural cavity which taps the underground water and into which surface waters, waste waters, industrial waste or sewage is placed.

Durham system An all-threaded pipe system of rigid construction, using recessed drainage fittings to correspond to the types of piping being used.

Effective opening The minimum cross-sectional area of the diameter of a circle at the point of water supply discharge.

Effluent The liquid waste as it flows from the septic tank and into the drainfield.

Fixture branch The drain from the trap of a fixture to the junction of that drain with a vent. Some codes refer to a fixture branch as a "fixture drain."

Fixture drain The drain from the fixture branch to the junction of any other drain pipe, referred to in some codes as a "fixture branch."

Fixture unit A design factor to determine the load-producing value of the different plumbing fixtures. For instance, the unit flow rate from fixtures is determined to be one cubic foot, or 7.5 gallons of water per minute.

Fire lines The complete wet standpipe system of the building, including the water service, standpipe, roof manifold, Siamese connections and pumps.

Flood level rim The top edge of a plumbing fixture or other receptacle from which water or other liquids will overflow.

Floor drain An opening or receptacle located at approximately floor level connected to a trap to receive the discharge from indirect waste and floor drainage.

Floor sink An opening or receptacle usually made of enameled cast iron located at approximately floor level which is connected to a trap, to receive the discharge from indirect waste and floor drainage. A floor sink is more sanitary and easier to clean than a regular floor drain, and is usually used for restaurant and hospital installations.

Flushometer valve A device actuated by direct water pressure which discharges a predetermined quantity of water to fixtures for flushing purposes.

Grade The slope or pitch, known as "the fall," usually expressed in drainage piping as a fraction of an inch per foot.

Horizontal pipe Any pipe or fitting which makes an angle of more than 45 degrees with the vertical.

Horizontal branch A drain pipe extending laterally from a soil or waste stack or building drain. May or may not have vertical sections or branches.

Indirect wastes A waste pipe charged to convey liquid wastes (other than body wastes) by discharging them into an open plumbing fixture or receptacle such as floor drain or floor sink. The overflow point of such fixture or receptacle is at a lower elevation than the item drained.

Industrial waste Liquid waste, free of body waste, resulting from the processes used in industrial establishments.

Insanitary Contrary to sanitary principles, injurious to health.

Interceptor A device designed and installed to separate and retain deleterious, hazardous, or undesirable matter from normal wastes and permit normal sewage or liquid wastes to discharge by gravity into the disposal terminal or sewer.

Leader The vertical water conductor or downspout from the roof to the building storm drain, combined building sewer, or other approved means of disposal.

Liquid waste The discharge from any fixture, appliance or appurtenance that connects to a plumbing system which does not receive body waste.

Load factor The percentage of the total connected fixture unit flow rate which is likely to occur at any point with the probability factor of simultaneous use. It varies with the type of occupancy, the total flow unit above this point being considered.

Loop or circuit waste and vent A combination of plumbing fixtures on the same floor level in the same or adjacent rooms connected to a common horizontal branch soil or waste pipe.

Main The principal artery of *any system* of continuous piping, to which branches may be connected.

Main vent The principal artery of the venting system, to which vent branches may be connected.

May The word "may" as used in the code book is a term of permission.

Mezzanine An intermediate floor placed in any story or room. When the total area of any such mezzanine floor exceeds 33⅓ percent of the total floor area in that room or story, it is considered an additional story rather than a mezzanine.

Pitch "Grade," also referred to as "slope."

Plumbing Includes any or all of the following: (1) the materials including pipe, fittings, valves, fixtures and appliances attached to and a part of a system for the purpose of creating and maintaining sanitary conditions in buildings, camps and swimming pools on private property where people live, work, play, assemble or travel; (2) that part of a water supply and sewage and drainage system extending from either the public water supply mains or private water supply to the public sanitary, storm or combined sanitary and storm sewers, or to a private sewage disposal plant, septic tank, disposal field, pit, box filter bed or any other receptacle or into any natural or artificial body of water or watercourse upon public or private property; (3) the design, installation or contracting for installation, removal and replacement, repair or remodeling of all or any part of the materials, appurtenances or devices attached to and forming a part of a plumbing system, including the installation of any fixture, appurtenance or devices used for cooking, washing, drinking, cleaning, fire fighting, mechanical or manufacturing purposes.

Plumbing fixtures Receptacles, devices, or appliances which are supplied with water or which receive or discharge liquids or liquid borne waste, with or without discharge, into the drainage system with which they may be directly or indirectly connected.

Plumbing official inspector The chief administrative officer charged with the administration, enforcement and application of the plumbing code and all its amendments.

Plumbing system The drainage system, water supply, water supply distribution pipes, plumbing fixtures, traps, soil pipes, waste pipes, vent pipes, building drains, building sewers, building storm drain, building storm sewer, liquid waste piping, water treating, water using equipment, sewage treatment, sewage treatment equipment, fire standpipes, fire sprinklers and related appliances and appurtenances, including their respective connections and devices, within the private property lines of a premises.

Potable water Water that is satisfactory for drinking, culinary and domestic purposes and meets the requirements of the health authority having jurisdiction.

Private property For the purposes of the code, all property except streets or roads dedicated to the public, and easements (excluding easements between private parties).

Private or *private use* In relation to plumbing fixtures: in residences and apartments, and in private bathrooms of hotels and similar installations where the fixtures are intended for the use of a family or an individual.

Private sewer A sewer privately owned and not directly controlled by public authority.

Public or *public use* In relation to plumbing fixtures: in commercial and industrial establishments, in restaurants, bars, public buildings, comfort stations, schools, gymnasiums, railroad stations or places to which the public is invited or which are frequented by the public without special permission or special invitation, and other installations (whether paid or free) where a number of fixtures are installed so that their use is similarly unrestricted.

Public sewer A common sewer directly controlled by public authority.

Public swimming pool A pool together with its buildings and appurtenances where the public is allowed to bathe or which is open to the public for bathing purposes by consent of the owner.

Relief vent A vent, the primary function of which is to provide circulation of air between drainage and vent systems.

Rim In code usage, an unobstructed open edge at the overflow point of a fixture.

Rock drainfield Three-quarter inch drainfield rock 100 percent passing a one inch screen and a maximum of ten percent passing a one-half inch screen.

Roof drain An outlet installed to receive water collecting on the surface of a roof which discharges into the leader or downspout.

Roughing-in The installation of all parts of the plumbing system that can be completed prior to the installation of plumbing fixtures; includes drainage, water supply, vent piping, and the necessary fixture supports.

Sanitary sewer A pipe which carries sewage and excludes storm, surface and ground water.

Second hand A term applied to material or plumbing equipment which has been installed and used, or removed.

Septic tank A watertight receptacle which receives the discharge of a drainage system or part thereof, so designed and constructed as to separate solids from liquid, digest organic matter through a period of detention, and allow the liquids to discharge into the soil outside the tank through a sub-surface system of open-joint or perforated piping, or other approved methods.

Sewage Any liquid waste containing animal, mineral or vegetable matter in suspension or solution. May include liquids containing chemicals in solution.

Shall As used in the code, a term meaning that compliance is mandatory and that the procedure or condition specified must be performed without deviation.

Slope See Grade.

Soil pipe Any pipe which conveys the discharge of water closets or fixtures having similar functions, with or without the discharge from other fixtures, to the building drain or building sewer.

Stack The vertical pipe of a system of soil, waste or vent piping.

Stack vent (Sometimes called a waste vent or soil vent) the extension of a soil or waste stack above the highest horizontal drain connected to the stack.

Standpipe system A system of piping installed for fire protection purposes having a primary water supply constantly or automatically available at each hose outlet.

Subsurface drain A drain which receives only subsurface or seepage water and conveys it to a place of disposal.

Sump A tank or pit which receives sewage or liquid waste, located below the normal grade of the gravity system and which must be emptied by mechanical means.

Supports (Also known as "hangers" or "anchors.") Devices for supporting and securing pipe and fixtures to walls, ceilings, floors or structural members.

Supply well Any artificial opening in the ground designed to conduct water from a source bed through the surface when water from such well is used for public, semi-public or private use.

Trap A fitting or device so designed and constructed as to provide a liquid seal which will prevent the back passage of air without materially affecting the flow of sewage or waste water through it.

Trap seal The maximum vertical depth of liquid that a trap will retain, measured between the crown weir and the top of the dip of the trap.

Vent stack A vertical vent pipe installed primarily for the purpose of providing circulation of air to and from any part of the drainage system.

Vent system A pipe or pipes installed to provide a flow of air to or from a drainage system or to provide a circulation of air within such system.

Vertical pipe Any pipe or fitting which is installed in a vertical position or which makes an angle of not more than 45 degrees with the vertical.

Waste pipe Any pipe which receives the discharge of any fixture, except water closets or fixtures having similar functions, and conveys it to the building drain or to the soil or waste stack.

Water-distributing pipe A pipe which conveys water from the water service pipe to the plumbing fixtures, appliances and other water outlets.

Water main A water supply pipe for public or community use.

Water outlet As used in connection with the water-distributing system, the discharge opening for the water (1) to a fixture, (2) to atmospheric pressure (except into an open tank which is part of the water supply system), (3) to a boiler or heating system, (4) to any water-operated device or equipment requiring water to operate, but not a part of the plumbing system.

Water service pipe The pipe from the water main or other source of water supply to the building served.

Water supply system Consists of the water service pipe, the water-distributing pipes, standpipe system and the necessary connecting pipes, fittings, control valves and all appurtenances in or on private property.

Wet vent A waste pipe which serves to vent and convey waste from fixtures other than water closets.

Yoke vent A pipe connecting upward from a soil or waste stack for the purpose of preventing pressure changes in the stacks.

E

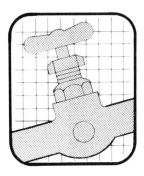

Standard Abbreviations

The abbreviations here are often found on blueprints (building plans) and in plumbing reference books (including the code) to identify plumbing fixtures, pipes, valves and nationally-recognized associations.

A	area	cu. in.	cubic inches
AD	area drain	C. W. M.	clothes washing machine
AGA	American Gas Association	C. V.	check valve
AISI	American Iron and Steel Institute	D. F.	drinking fountain
ASA	American Standard Association	D. W.	dish washer
ASCE	American Society of Civil Engineering	E to C	end to center
ASHRAE	American Society of Heating, Refrigeration and Air Conditioning Engineers	EWC	electric water cooler
		°F	degrees Fahrenheit
		F	Fahrenheit
ASME	American Society of Mechanical Engineers	F. B.	foot bath
		F. F.	finish floor
ASSE	American Society of Sanitary Engineering	F. C. O.	floor cleanout
		F. D.	floor drain
ASTM	American Society for Testing Materials	F. D. C.	fire department connection
		F. E. C.	fire extinguisher cabinet
AWWA	American Water Works Association	F. G.	finish grade
B.S.	bar sink	F. H. C.	fire hose cabinet
B	bidet	F. L.	fire line
B. T.	bathtub	F. P.	fire plug
B.t.u.	British Thermal Unit	F. S. P.	fire standpipe
C to C	center to center	F. U.	fixture unit
CI	cast iron	GAL.	gallons
CISPI	Cast Iron Soil Pipe Institute	gpm or Gal. per Min.	gallons per minute
C	condensate line		
C. O.	cleanout		
C. W.	cold water		
cu. ft.	cubic feet	Galv.	galvanized

G. S.	glass sink	P. D.	planter drain
G. V.	gate valve	P. P.	pool piping
G. P. D.	gallons per day	PSI	pounds per square inch
H. B.	hose bibb	Rad.	radius
Hd or H.D.	head	R. D.	roof drain
H. W.	hot water	red.	reducer
H. W. R.	hot water return	R. L.	roof leader
HWT	hot water tank	San.	sanitary
in.	inch	SH.	Shower
I. D.	inside diameter	Spec.	specification
I W	indirect waste	Sq.	square
IPS	iron pipe size	S. B.	Sitz bath
K. S.	kitchen sink	Sq. Ft.	square feet
L. or LAV.	lavatory	S. P.	swimming pool
L. T.	laundry tray	SS	service sink
L	length	Std.	Standard
lb.	pound	SV	service
Max.	maximum	SW	service weight
Mfr.	manufacturer	S & W	soil and waste
Min.	minimum	T	temperature
M. H.	manhole	U or Urn	urinal
NAPHCC	National Association of Plumbing Heating and Cooling Contractors	V	volume
NBFU	National Board of Fire Underwriters	Vtr	vent through roof
NBS	National Bureau of Standards	W	waste
NFPA	National Fire Protection Association	W.C.	water closet
NPS	nominal pipe size	W.H.	water heater
O	oxygen	XH	extra heavy
O. D.	outside diameter		
Oz.	ounce		

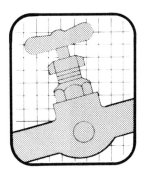

Common Fixture, Plumbing, & Fitting Symbols

Fixture Symbols

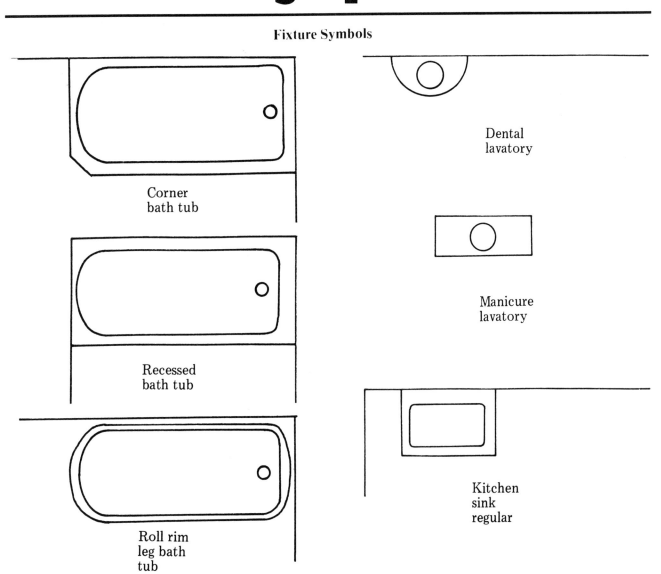

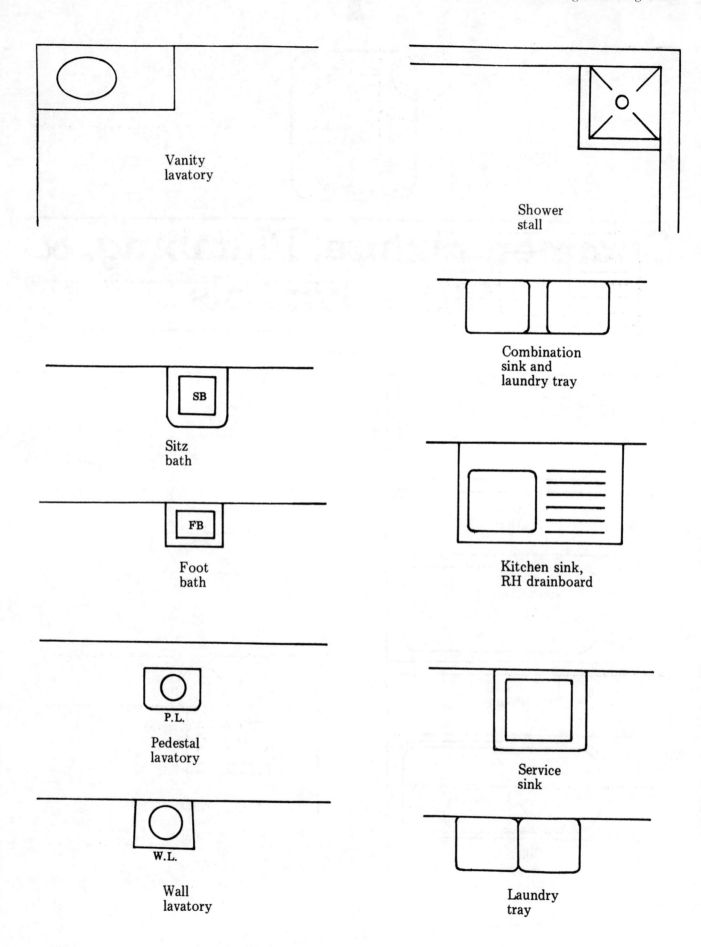

Common Fixture, Plumbing, & Fitting Symbols

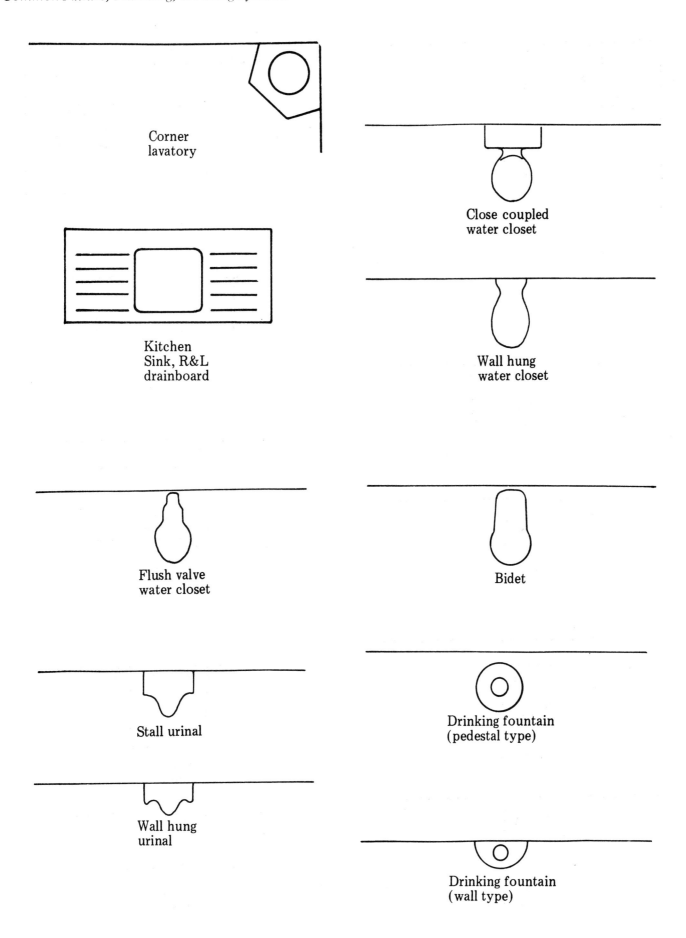

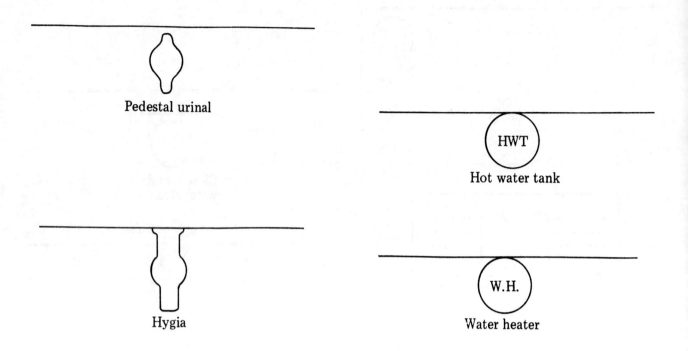

Common Fixture, Plumbing, & Fitting Symbols

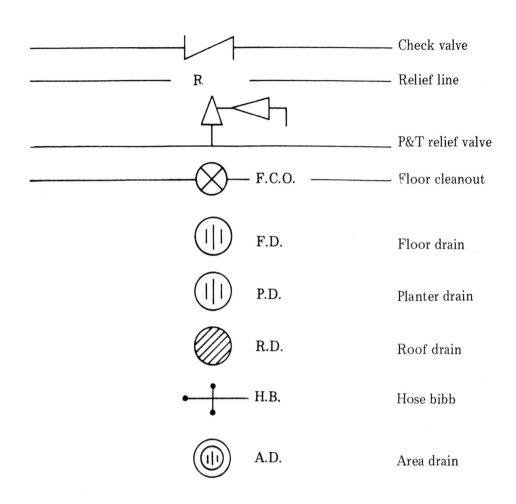

FITTING SYMBOLS

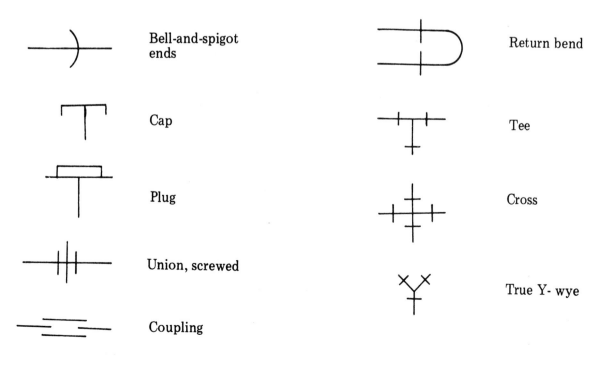

Expansion joint, sliding	Y- Wye single
Sleeve	Combination wye and 1/8 bend
Expansion joint, bellows	Tee union
Bushing	Y- Wye double
Reducer	Double combination wye and 1/8 bend
Eccentric reducer	
Reducing flange	Elbow, 90 degrees
Union, flanged	Tee, outlet up
Elbow, 45 degrees	Tee, side outlet up
Elbow, 16 degrees	
Double ¼ bend	Screwed ends

Common Fixture, Plumbing, & Fitting Symbols

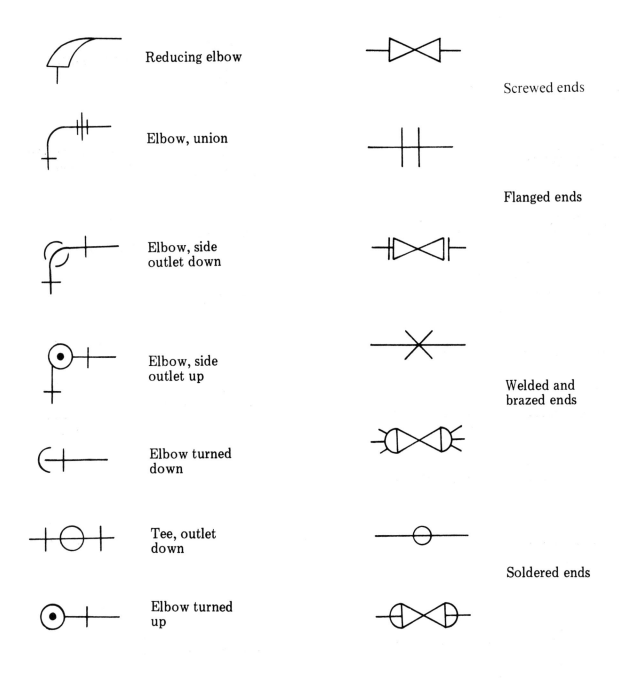

Index

A
Abbreviated symbols, fixtures 14-15
Abbreviations 211-212
Above-ground fire lines 60, 61, 63, 64
Above-ground plot plans 8, 25, 30
Acceptable manufacturers 176
Accessory building hook-up 99, 100
Accounting and estimating 115
Accuracy, estimating materials 123
Accurate, detailed estimate 123
Acid neutralizing tanks 168
Addendum 123
Additional compensation 120
Additional fixture requirements 190-191
Additional time requirements 124
Adjustable fire hose nozzle 61
Air conditioning condensate drain 49, 119
American Institute of Architects 121
Anticipatory breach 120
Approved job plans 8
Architect's floor plan 16-24, 26-27, 31
Architect's status 121
Architectural scale rule 133
Area, standard pipe 197
As-built plans 8
Assess damages 119
Auxiliary alarm panel 176

B
Backfill, trailer parks 57
Backflow preventer 168
Basic isometric angles 35
Bathtub 167
Bedding, trailer parks 57
Begin take-off 133
Below-ground plot plans 8, 25, 30
Bid exclusions 120
Bid price total 132
Bids 111
Binding contract 119, 120
Blueprints 123
Bottled gas 69
Breach of contract 120

Building piping systems 159
Building sewer pipes 75, 76
Business failures 114
Butane 69

C
Calculating fixture requirements 190
Calculating sewage flow 190
Calculating water supply systems 183-190
Cancellation of contracts 119-120
Cancelling the contract 120
Capacity of water pipes 187-190
Catch basins 9, 25, 30
Changes in contract 120-121
Changes in specifications 120-121
Changes, written addendum 123
Checklist 149-152
Circle, terms and calculations 197-200
Circulating pumps 168
Circumference 197
Classes of drainage pipes and fittings 41
Cleaning water piping installation 177
Cleanouts 182
Code 123
Code specifications 174
Cold water systems 137
Combination waste and vent system 16, 96, 102
Commercial baths 168
Commercial building 9
Commercial gas example 71
Commercial job specifications 174-182
Commercial plumbing fixtures 167-169
Commercial sinks 167-168
Commercial water service pipe 185
Common pipe offset 198
Compensation 119, 120
Compiling the estimate 155-157
Completion problems 120
Complex specifications 123
Compression joints 160-161, 164
Condensate drain, air conditioning 49
Connecting parallel piping 198
Connecting sewers 75-76

Constant pressure pumps 168
Construction expenses 116
Contingency cost 155
Contract, act of omission 121
Contract, altered conditions 121
Contract, binding 119, 120
Contract, lawful considerations 119, 120
Contract, legally binding 157
Contract, "literal" performance 121
Contract, "substantial" performance 121
Contract cancellation 119, 120
Contract changes 120-121
Contract considerations 117
Contract language 120
Contract requirements 122
Contractor, estimator 110-112
Contractor, general 112
Contractor's limitations 111
Contractor's over-expansion 111
Contractor's responsibility piping system test 177
Converting minutes to decimal hours 169
Coordination specifications 174
Corrosive liquids 47
Corrosive waste piping 48
Cost of changes 120-121
"Cost plus" contracts 119
"Cost plus" jobs 115
Cost sheet, material quantity 16
Cross connection, prevention 48

D
Daily sewage flow 194-195
Damage assessment 119
Deck drains 181
Dependent trailer 55
Design, material take-off 16-24, 28-29, 32-33
Detailed estimate 114
Detailed material take-off 123
Diameter of circle 197
Dielectric unions installation 177
Dilution tanks 47
Document, American Institute of Architects 121
Domestic water 178-179
Domestic water system 176

Index

Drainage pipe 41
Drainage waste/vent system 133, 137
Drainage well detail 94, 97
Drawings specifications 174
Drinking fountain 168
Dry gas 69

E

Electric water heater 179
Elements of a contract 119
Emergency connection riser detail 90
Equipment quantity cost sheet 145
Equipment requirement specifications 174
Erroneous bid 120
Estimate, labor 141
Estimate maximum demand 183-184
Estimate sample 125
Estimating and accounting 115
Estimating forms 126-132
Estimating gas system 69
Estimating materials 123
Estimating trailer park systems 57
Estimating variations 114
Estimator, contractor 110-112
Everyday estimating problems 8
Excavation, piping system 177
Exclude from your bid 120
Expenses, construction 115-116
Exterior domestic water 178
Extra charges 120, 121
"Extras" 120-121

F

Factor of simultaneous use 186
Faulty work 121
Fees, piping system 176
Fire, auxiliary alarm panel 176
Fire connection, accessibility 60
Fire connection, roof manifold 60-63
Fire connection, siamese 63
Fire connection, water main 60
Fire department connections 59, 63-65
Fire detector check valve 60
Fire extinguisher 176
Fire flow, on-site well system 66
Fire hose, wall-mounted 60
Fire hose adjustable nozzle 61
Fire hose outlets 60
Fire hydrant installation 66-68
Fire hydrants 67
Fire inspection test valve 64
Fire jockey pumps 176
Fire lines 59-64
Fire National Standard Thread 60-63
Fire post indicator valve 60
Fire pressure reducer 61
Fire protection, supplementary documents 175
Fire protection systems 59-68, 168, 175-176
Fire pump controllers 175-176
Fire pump detail 61
Fire pump planning regulations 65
Fire pump system 175-176
Fire pump trim 175
Fire pumps 61-65
Fire standpipes 59-65
Fire system bypass 65
Fire system jockey pump 65
Fire Underwriters' Laboratory 60-63
Fire Underwriters' listed check valve 63
Fire wall-mounted hose reel 60
Fire well standard detail 66
Fitting classes 41
Fitting quantity cost sheets 136, 138, 141
Fitting symbols 217-219
Fittings, isometric 16
Fixture carrier 182
Fixture clearances 14-15
Fixture clearances, handicapped 15
Fixture pipe size 201
Fixture quantity cost sheet 144
Fixture requirements 190-191
Fixture spacing 14-15
Fixture supply pipe, size 183
Fixture supports 182
Fixture symbols 14-15, 213-216
Fixture unit value 187
Fixtures, public assembly 191-193

Flap check valve 49
Flare joints 164
Flat wet vent system 16-23
Floor drainage, special 50
Floor drains 44, 181
Floor plans 14, 17-24, 26-27, 31
Floor plans, to scale 124
Floor-mounted interceptors 44
Flow pressure 200-201
Follow contract 120-121
Follow specifications 120-121
Foreman 115, 116
Form, subcontract agreement 118

G

Gas appliance manufacturer 70
Gas Btu's per hour 70-73
Gas, butane 69
Gas, chemical odorant 69
Gas, commercial installation 71
Gas, cubic feet 70-73
Gas, demand maximum 70-73
Gas, developed length 71
Gas, dry 69
Gas, types 69
Gas, liquified petroleum 69
Gas, low pressure system 70-73
Gas, LP or bottled 69
Gas, manufactured 69
Gas, methane 69
Gas, natural 69
Gas, pipe size rule 70
Gas, pressure loss 70
Gas, propane 69
Gas, residential installation 72-73
Gas, specific gravity 70
Gas, sweet 69
Gas system, estimating 69
Gas system, sized 70-73
Gas systems 69-73
Gasoline interceptors 44-45
General and subcontractor 122
General contractor 112, 122
General estimating outline 124
General summary check sheet 149-150
Good estimate 113, 114
Good job management 112
Government policy 8
Grease interceptor 43-48, 168
Grease trap, diagram 44
Guarantee, residential 174

H

Handicapped, fixtures and facilities 14-16
Hand excavation 166
Heads of water (psi) 200
Horizontal pipe hangers 166, 170
Horizontal storm drain 51
Horizontal stormwater pipes 53
Horizontal wet vents 96, 101
Hose outlets, fire 60
Hot water generator 168
Hot water heater 167-168
Hot water systems 137, 140
Hydrants, domestic 182
Hydropneumatic system, water 176
Hydrostatic table 200

I

Independent trailer 55
Indirect waste piping installation 48-49
Indirect waste system 45
Individual vent system 16, 24
Inflation 8
Inside sanitary piping 178
Inspection 7
Install valve 177
Installation, cleaning water piping 177
Installation, dielectric unions 177
Installation, pipe supports 177
Installation, sleeves 177
Installation methods, piping system 177
Interceptors 43-48
Interior domestic water 179
Interior drainage piping 162
Isometric, drainage and vent 41, 76, 101-104
Isometric, fittings 16, 38-40

Isometric, material take-off 17-24, 28-29, 32-33
Isometric, plumbing design 16-24, 28-29, 32-33
Isometric, right triangle 37
Isometric, type and number fittings 16-40
Isometric, various angles 36
Isometric angles 35-36
Isometric drawings, instructions, 35-41
Isometric flat wet vent system 17-23
Isometric individual vent system 24
Isometric stacked system 17-24, 29
Itemizing costs 119

J

Job management 111, 112
Job overhead expense 115, 155
Job plans 8
Job site 8, 77
Job site expenses 116
Job specifications 8
Jockey pump controller 176
Jockey pumps 176
Joints, compression 160-161, 164
Joints, flare 164
Joints, lead and oakum 137, 160, 162
Joints, no-hub 160, 162
Joints, soldered 161-162, 164
Joints, solvent cemented 161-162, 164
Joints, tapered friction 161
Joints, threaded 164-165

K

Kitchen sink 167

L

Labor cost sheets 130, 132
Labor cost summary sheets 146-148
Labor estimates 140
Laundry interceptors 45
Laundry sink 167
Lavatory 167
Lavatory, handicapped 15
Layouts 14, 16, 28-29, 32-33
Layouts, sample drawings 16-33
Lead, oakum, gas joints 137
Liens, release 119
Linear feet, piping 133, 135, 138-139
Lint interceptors 46-47
Liquidated damages 119
Liquified petroleum gas 69
Local utility company 8
Logical sequence 8

M

Mandates of plumbing code 7
Manhole detail, diagrams 78-85
Manhole flow patterns 80
Manhole frame and cover 84, 85
Manhole requirements 75, 77
Manhole section 78-79, 81-83
Man-hour production tables 159-169
Man-hour tables, use 124-125, 159
Man-hours, plumbing fixtures 166-168
Manufactured gas 69
Material, valves 177-178
Material discounts 141
Material quantity cost sheet 16
Material take-off 123-125
Material take-off, design 16-24, 28-29, 32-33
Materials and methods 175-176
Maximum fixture demand 188, 190
Maximum rainfall rate 51
Measuring offsets 199-200
Mechanic's lien laws 122
Meter box 179
Methane gas 69
Minutes to decimal hours 169
Miscellaneous quantity cost sheet 143

N

Natural gas 69
Neutralizing tanks 47
No-hub joints 160, 162
Noise specifications 174
Non-working foreman 115, 116

O

Office overhead expense 114, 115, 155

Offset, pipe 198
Oil interceptors 44-45, 168
Oil interceptors, capacity 45
Oil interceptors, material 45
On-site fire systems 68
On-site fire well 65
Oral contracts 119
Oral understanding 117
Other drainage systems 137
Outside sanitary sewer 178
Overhead expense, job 115
Overhead expense, office 114, 115
Overhead, real cost 119

P
Parallel piping 198
Partial payments 119
Peak demand 184
Penalties for non-performance 119
Penalty and indemnity clauses 119
Percentage ratio 191
Permit, sewer connection 75
Pipe equivalents 198
Pipe insulation 165
Pipe painting 165
Pipe sleeves 165
Pipe support, installation 177
Piping, "rule of thumb" 140
Piping installations 162
Piping quantity cost sheets 135, 138, 139
Piping system, excavation 177
Piping system, installation 177
Piping systems, fees 176
Piping systems, general requirements 176
Piping systems, utilities 176
Planning regulations, fire pump 65
Plans, "as-built" 8
Plans, job approved 8
Plot plans 8-12, 25, 30, 77
Plumbing code mandates 7
Plumbing design isometrics 16-24, 28-29, 32-33
Plumbing drains and specialties 181
Plumbing equipment 140, 179
Plumbing estimator 113-116
Plumbing fixtures 141, 180-181
Plumbing fixtures, abbreviations 14-15
Plumbing mathematics 197-200
Plumbing permits 116, 155
Plumbing symbol legend 216-217
Plumbing symbols 213-219
Plumbing system 7
Plumbing systems, multi-story 96, 103, 104
Pressure in pounds 200
Pressure piping 163
Pricing the estimate 133-135
Private sewer, connection 75, 76
Probable demand 187-188
Procedure for estimate 133-153
Project summary sheet 131-132, 155-156
Profit 155
Propane gas 69
Public assembly fixtures 191-193
Public sewer, connection 75, 76
Public water 9, 25, 30
Pump components, water system 176
Pump operation, water system 176

Q
Quantity cost sheets 128-129

R
Radius 197
Rainfall 51
Rate of water flow 200
Recapitulation sheet 156, 157
Record drawings specification 174
Required water pressure 200
Requirements, special 8
Reseal, trap detail 93, 96
Residential gas example 73
Residential installation, gas 72
Residential plot plans 8-11
Residential plumbing fixtures 167
Residential water service pipe 184
Revised plot plans 8
Right to stop work 120
Roof area, stormwater 51, 52

Roof drains 181
Roof slope, stormwater 51
Rule, gas pipe size 70
Rule, maximum rainfall rate 51

S
Sales tax 116, 155
Sample drawings, layouts 16-33
Sample estimate 135-136, 138-139, 141-148
Sample standard analysis sheet 127
Sand interceptors 44-45
Sanitary piping 178
Sanitary sewer detail, diagrams 78-87
Sanitary sitework systems 75-104
Scale of floor plans 124
Scope specifications 174
Selling price 157
Septic tanks 8
Service connection, trailers 58
Settling tank 94, 97
Sewage collection system 77
Sewage ejector pumps 168
Sewage ejectors 76, 93
Sewage flow 190, 194-195
Sewage force main connection 91
Sewage force mains 76, 88, 90-92
Sewage lift stations 76, 77, 88-90
Sewer, accessory building 96
Sewer, shallow connection 86
Sewer, typical riser connection 87
Sewer connection 9, 25, 30
Sewer connection permit 75
Sewer pipes 75-76
"Sharp" operator 117
Shortages 8
Shower 167
Siamese fire connection 63
Side agreements 117
Simple job specifications 173-174
Simple sanitary system 41
Simple specifications 123
Site 8, 77
Sitework 8
Sitework, building sewer 75, 77, 96, 99, 100
Sitework, water distribution 105, 106
Size piping 133-134
Sizing gas systems 70-73
Sizing stormwater drains 51
Sizing trailer park drainage 56
Sizing waste piping 48, 49
Sizing water service pipe 184
Sleeve installation 177
Soakage pit 9, 25, 30
Solder 140
Spacing, fixtures 14-15
Special fixtures 187
Special fixtures and connection 167
Special requirements 8
Special waste interceptors 43
Special waste piping 48-54
Specialty contractors 8
Specification changes 120-121
Specifications 123
Specified forfeiture 119
Sprinkler system 65
Square footage/occupant 193
Stacked system 16-24, 29
Stacked wet vent system 16
Standard analysis sheet 127
Standard contract form 118
Standard take-off forms 126
Standpipes 59-65, 175
Statute of frauds 119
Storm drain, horizontal 51
Storm drain, sizing 49-54
Storm drainage, rainfall rate 53-54
Storm drainage system 49-54
Stormwater, roof area 51, 52
Stormwater, roof slope 51
Stormwater, vertical wall 51
Stormwater drains, sizing 51
Subcontract agreement 117-122
Subcontractor's bid 8
Submittal specifications 174
Sub's job cost 8
Substitution specifications 174
Suing rights 120

Sump pump 76, 93, 94, 168
Superintendent 115
Supplementary costs 155
Supplementary documents, fire 175
Supports 170
Sweet gas 69
Symbol plumbing legend 216-217
Symbols, fitting 217-219
Symbols, fixture 213-216
Symbols, plumbing 213-219

T
Take-off 133-153
Tanks, dilution 47
Tanks, neutralizing 47
Tapered friction joints 161
Temporary expenses 116
Temporary water 155
Testing piping system 177
Thrust blocks 60, 62, 92
Time and material 119
Time requirements 124
Toilet rooms, handicapped 14-16
Trailer coach, independent 55
Trailer park cleanouts 57, 58
Trailer park connection 58
Trailer park estimating 57
Trailer park fixture units 56
Trailer park sanitary facilities 55
Trailer park sewage collection 56-57
Trailer park vents 57
Trailer parks 55-58
Transformer oil holding tank 95
Transformer vault drainage 93, 95
Trap reseal detail 93, 96
Traps 41
Trash chute piping 94, 98
Trash room drainage 94, 98
Trash room fire sprinkler 94, 98
Travel trailer, dependent 55

U
Underground fire lines 59-60
Understanding your contract 121
Unreasonably delayed work 120
Urinal 168
Urinals, handicapped 15-16
Using man-hour tables 124-125
Utilities, piping system 176
Utility company 8

V
Vacuum breakers 182
Valves 177-178
Valves quantity cost sheet 142
Vents 41
Vertical piping supports 166, 170
Vertical wall, stormwater 51
Volume, tanks 199

W
Waste and vent system 16, 96, 102
Water closet 167-168
Water closet, handicapped 14
Water coolers, handicapped 16
Water distribution systems 105-109
Water heater, electric 179
Water heater components 179
Water main air release valve 109
Water main valve setting 108
Water meter bank detail 107
Water meter box 179
Water meter cost 106
Water meter installation 105-107
Water meters 9, 25, 30, 178-179
Water pipe sizing 185-190
Water service pipe 184-185
Water system pumps 176
Wells 8
Working foreman 115, 116
Writing up contract 157
Written addendum changes 123
Written agreement 117
Written specifications 120-121

Y
Yard and street fire hydrants 68

Practical References For Builders

Plumbers Handbook
Explains in simple terms and shows with clear drawings what will pass inspection: vents, waste piping, drainage, septic tanks, hot and cold water supply systems, wells, fire protection piping, fixtures, solar energy systems, gas piping, and more. Based on the current edition of the popular national code that is standard for most cities and code jurisdictions. Many pages of practical recommendations and trade tips are included: common layouts for residences, how to size piping, preparing for the journeyman's exam. **224 pages, 8½ x 11, $13.50**

Basic Plumbing With Illustrations
The journeyman's and apprentice's guide to installing plumbing, piping and fixtures in residential and light commercial buildings: how to select the right materials, lay out the job and do professional quality plumbing work. Explains the use of essential tools and materials, how to make repairs, maintain plumbing systems, install fixtures and add to existing systems. **320 pages, 8½ x 11, $15.50**

National Construction Estimator
Current building costs in dollars and cents for residential, commercial and industrial construction. Prices for every commonly used building material, and the proper labor cost associated with installation of the material. Everything figured out to give you the "in place" cost in seconds. Many time-saving rules of thumb, waste and coverage factors and estimating tables are included. **480 pages, 8½ x 11, $14.75. Revised annually.**

Planning and Designing Plumbing Systems
Explains in clear language, with detailed illustrations, basic drafting principles for plumbing construction needs. Covers basic drafting fundamentals: isometric pipe drawing, sectional drawings and details, how to use a plot plan, and how to convert it into a working drawing. Gives instructions and examples for water supply systems, drainage and venting, pipe, valves and fixtures, and has a special section covering heating systems, refrigeration, gas, oil, and compressed air piping, storm, roof and building drains, fire hydrants, and more. **8½ x 11, 224 pages, $13.00**

Masonry & Concrete Construction
Every aspect of masonry construction is covered, from laying out the building with a transit to constructing chimneys and fireplaces. Explains footing construction, building foundations, laying out a block wall, reinforcing masonry, pouring slabs and sidewalks, coloring concrete, selecting and maintaining forms, using the Jahn Forming System and steel ply forms, and much more. Everything is clearly explained with dozens of photos, illustrations, charts and tables. **224 pages, 8½ x 11, $13.50**

Building Cost Manual
Square foot costs for residential, commercial, industrial, and farm buildings. In a few minutes you work up a reliable budget estimate based on the actual materials and design features, area, shape, wall height, number of floors and support requirements. Most important, you include all the important variables that can make any building unique from a cost standpoint. **240 pages, 8½ x 11, $10.00. Revised annually**

Estimating Home Building Costs
Estimate every phase of residential construction from site costs to the profit margin you should include in your bid. Shows how to keep track of manhours and make accurate labor cost estimates for footings, foundations, framing and sheathing finishes, electrical, plumbing and more. Explains the work being estimated and provides sample cost estimate worksheets with complete instructions for each job phase. **320 pages, 5½ x 8½, $14.00**

Construction Estimating Reference Data
Collected in this single volume are the building estimator's 300 most useful estimating reference tables. Labor requirements for nearly every type of construction are included: site work, concrete work, masonry, steel, carpentry, thermal & moisture protection, doors and windows, finishes, mechanical and electrical. Each section explains in detail the work being estimated and gives the appropriate crew size and equipment needed. Many pages of illustrations, estimating pointers and explanations of the work being estimated are also included. This is an essential reference for every professional construction estimator. **368 pages, 11 x 8½, $18.00**

Concrete Construction & Estimating
Explains how to estimate the quantity of labor and materials needed, plan the job, erect fiberglass, steel, or prefabricated forms, install shores and scaffolding, handle the concrete into place, set joints, finish and cure the concrete. Every builder who works with concrete should have the reference data, cost estimates, and examples in this practical reference. **571 pages, 5½ x 8½, $17.75**

Wood Frame House Construction
From the layout of the outer walls, excavation and formwork, to finish carpentry, and painting, every step of construction is covered in detail with clear illustrations and explanations. Everything the builder needs to know about framing, roofing, siding, insulation and vapor barrier, interior finishing, floor coverings, and stairs...complete step by step "how to" information on what goes into building a frame house. **240 pages, 8½ x 11, $9.75. Revised edition**

National Repair And Remodeling Estimator
The complete pricing guide for dwelling reconstruction costs. Reliable, specific data you can apply on every remodeling job. Up-to-date material costs and labor figures based on thousands of repair and remodeling jobs across the country. Professional estimating techniques to help determine the material needed, the quantity to order, the labor required, the correct crew size and the actual labor cost for your area. **240 pages, 8½ x 11, $15.25. Revised annually**

Building Layout
Shows how to use a transit to locate the building on the lot correctly, plan proper grades with minimum excavation, find utility lines and easements, establish correct elevations, lay out accurate foundations and set correct floor heights. Explains planning sewer connections, leveling a foundation out of level, using a story pole and batter boards, working on steep sites, and minimizing excavation costs. **240 pages, 5½ x 8½, $11.75**

Builder's Guide to Construction Financing
Explains how and where to borrow the money to buy land and build homes and apartments: conventional loan sources, loan brokers, private lenders, purchase money loans, and federally insured loans. How to shop for financing, get the valuation you need, help your buyers get their loan, comply with lending requirements, and handle liens. Includes chapters on tract, leasehold, condo, and special purpose loans. **304 pages, 5½ x 8½, $11.00**

Process & Industrial Pipe Estimating
A clear, concise guide to estimating costs of fabricating and installing underground and above ground piping. Includes types of pipes and fittings, valves, filters, strainers, and other in-line equipment commonly specified, and their installation methods. Shows how a take-off is consolidated on the estimate form and the bid estimate derived using the complete set of manhour tables provided in this complete manual of pipe estimating. **240 pages, 8½ x 11, $18.25**

Contractor's Guide To The Building Code
Explains in plain English exactly what the Uniform Building Code requires and shows how to design and construct residential and light commercial buildings that will pass inspection the first time. Suggests how to work with the inspector to minimize construction costs, what common building short cuts are likely to be cited, and where exceptions are granted. If you've ever had a problem with the code or tried to make sense of the Uniform Code Book, you'll appreciate this essential reference. **312 pages, 5½ x 8½, $16.25**

Builder's Office Manual
This manual will show every builder with from 3 to 25 employees the best ways to: organize the office space needed, establish an accurate record-keeping system, create procedures and forms that streamline work, control costs, hire and retain a productive staff, minimize overhead, shop for computer systems, and much more. Explains how to create routine ways of doing all the things that must be done in every construction office in a minimum of time, at lowest cost and with the least supervision possible. **208 pages, 8½ x 11, $13.25**

Contractor's Year-Round Tax Guide
How to set up and run your construction business to minimize taxes: corporate tax strategy and how to use it to your advantage, why you should consider incorporating to save tax dollars, and what you should be aware of in contracts with others. (Includes sample contracts). Covers tax shelters for builders, write-offs and investments that will reduce your taxes, accounting methods that are best for contractors, what forms of compensation are deductible, and what the I.R.S. allows and what it often questions. Explains how to keep records and protect your company from tax traps that many contractors fall into. **192 pages, 8½ x 11, $16.50**

Process Plant and Equipment Cost Estimation
Current cost data and estimating methods for process plant construction. Includes nearly 100 pages of cost data from a broad sample of U.S., European and Asian projects: typical equipment and plant costs, labor cost and productivity comparisons for the entire project duration, escalation indexes for both plant and equipment, manpower distributions, cost adjustments based on the selection of alternate materials, typical project durations and cost overruns, the cost of chemicals, construction materials and utilities, location cost indexes, and operating costs. **240 pages, 8½ x 11, $19.00**

Manual of Professional Remodeling
This is the practical manual of professional remodeling written by an experienced and successful remodeling contractor. Shows how to evaluate a job and avoid 30-minute jobs that take all day, what to fix and what to leave alone, and what to watch for in dealing with subcontractors. Includes chapters on calculating space requirements, repairing structural defects, remodeling kitchens, baths, walls and ceilings, doors and windows, floors, roofs, installing fireplaces and chimneys (including built-ins), skylights, and exterior siding. Includes blank forms, checklists, sample contracts, and proposals you can copy and use. **400 pages, 8½ x 11, $18.75**

Remodelers Handbook
The complete manual of home improvement contracting: Planning the job, estimating costs, doing the work, running your company and making profits. Pages of sample forms, contracts, documents, clear illustrations and examples. Chapters on evaluating the work, rehabilitation, kitchens, bathrooms, adding living area, re-flooring, re-siding, re-roofing, replacing windows and doors, installing new wall and ceiling cover, repainting, upgrading insulation, combating moisture damage, estimating, selling your services, and bookkeeping for remodelers. **416 pages, 8½ x 11, $18.50**

Finish Carpentry
The time-saving methods and proven shortcuts you need to do first class finish work on any job: cornices and rakes, gutters and downspouts, wood shingle roofing, asphalt, asbestos and built-up roofing, prefabricated windows, door bucks and frames, door trim, siding, wallboard, lath and plaster, stairs and railings, cabinets, joinery, and wood flooring. **192 pages, 8½ x 11, $10.50**

Spec Builder's Guide
Explains how to find the right lot in the right area and at the right price, get the financing you need, plan, build, and then sell the house at a price that earns a decent return on the time and money you've invested. Includes professional tips to ensure success as a spec builder: how to cut costs at every opportunity without sacrificing quality, avoiding losses, anticipating buyer preferences, and taking advantage of construction cycles. Every chapter includes checklists, diagrams, charts, figures, and estimating tables that make this an invaluable reference for every speculative or custom home builder. **448 pages. 8½ x 11, $24.00**

Craftsman BOOK COMPANY
6058 Corte del Cedro
P. O. Box 6500
Carlsbad, CA 92008

Mail Orders
We pay shipping when your check covers your order in full.

Name

Company

Address

City   State   Zip

Send check or money order
Total Enclosed _____ (In California add 6% tax)
If you prefer, use your ☐ Visa or ☐ MasterCard

Card no.

Expiration date _____ Initials _____

In a hurry?
We accept phone orders charged to your MasterCard or Visa. Call (619) 438-7828

10 Day Money Back GUARANTEE

☐ 15.50 Basic Plumbing with Illustrations
☐ 11.00 Builders Guide to Const. Financing
☐ 13.25 Builders Office Manual
☐ 10.00 Building Cost Manual
☐ 11.75 Building Layout
☐ 17.75 Concrete Const. & Estimating
☐ 18.00 Construction Estimating Reference Data
☐ 16.25 Contractor's Guide To The Building Code
☐ 16.50 Contractor's Year-Round Tax Guide
☐ 14.00 Estimating Home Building Costs
☐ 10.50 Finish Carpentry
☐ 18.75 Manual of Professional Remodeling
☐ 13.50 Masonry & Concrete Construction
☐ 14.75 National Construction Estimator
☐ 15.25 National Repair & Remod. Estimator
☐ 13.00 Planning & Designing Plumbing Systems
☐ 13.50 Plumbers Handbook
☐ 18.25 Process & Industrial Pipe Estimating
☐ 19.00 Process Plant & Equip. Cost Estimating
☐ 18.50 Remodelers Handbook
☐ 24.00 Spec Builders Guide
☐ 9.75 Wood-Frame House Construction
☐ 17.25 Estimating Plumbing Costs

These books are tax deductible when used to improve or maintain your professional skill.